"Praveen Gupta has developed an excellent methodology for aligning senior management, supervisors, and employees in continuously improving the critical drivers of sustained high performance."

> —Robert F. Anderson, President, Robert F. Anderson and Associates, Inc.

"This book must be studied by all who are driven to create a corporate environment that attracts good employees, demands results, and produces sustained profitability."

> —Lonnie C. Rogers, President, Ideal Aerosmith

"The Six Sigma Scorecard simplifies implementation of the Six Sigma initiative by using measurements for better, faster, and cost-effective performance throughout the company. The Business Performance Index (BPIn) is an excellent tool for executives to monitor corporate performance for growth and profitability."

> —Anton Hirsch, President, Cobra Metal Works

"The *Six Sigma Business Scorecard* makes it easy to understand the value of the process measurements by explaining their relationship to profitability. Using a business scorecard is an excellent idea to ensure improvement."

> —Chuck Georges, Vice President of Manufacturing, Molon Motor & Coil Corporation

"Synthesizing the tenets of Six Sigma methodology and the SIPOC elements with the balanced scorecard, Praveen has given leaders in all disciplines the business tools to measure and direct their effectiveness in terms of profitability and growth."

> —Brian O'Reilly, Director of Quality & Regulatory Affairs, Dentsply Professional

ABOUT THE AUTHOR

Praveen Gupta, President of Accelper Consulting, has gained extensive experience in the field of business management over 25 years. He has been a pioneer of Six Sigma methodology at Motorola, and of optimizing business performance by using his Execution Map consisting of Business Scorecard, Six Sigma, innovation, and process management. For each of the four elements, he has authored a book providing simple-to-use but powerful tools. Mr. Gupta writes a monthly column, "Six Sigma Sense," and has authored or co-authored several books, including *The Six Sigma Performance Handbook, Healthcare Excellence,* and *Business Innovation in the 21st Century.* He holds a BSEE from the Indian Institute of Technology and an MSEE from the Illinois Institute of Technology. He is also a member of the Chicago Executive Club, the Turnaround Management Association, and IPC. Mr. Gupta is a Master Six Sigma Black Belt and a Fellow of ASQ. He teaches at DePaul University and the Illinois Institute of Technology, both in Chicago, Illinois.

SIX SIGMA BUSINESS SCORECARD
Ensuring Performance for Profit

PRAVEEN GUPTA

SECOND EDITION

McGraw-Hill
New York • Chicago • San Francisco • Lisbon • London • Madrid
Mexico City • Milan • New Delhi • San Juan • Seoul
Singapore • Sydney • Toronto

The McGraw·Hill Companies

Library of Congress Cataloging-in-Publication Data

Gupta, Praveen, date.
  Six Sigma Business Scorecard : ensuring performance
for profit / Praveen Gupta.—2nd ed.
      p.   cm.
  Includes bibliographical references and index.
  ISBN 0-07-147943-0 (alk. paper)
  1. Six sigma (Quality control standard)   2. Quality control—
Statistical methods.   3. Production management—Statistical
methods.   I. Title.

  TS156.G868   2006
  658.4′013—dc22   2006033798

2 3 4 5 6 7 8 9 0   DOC/DOC   0 1 2 1 0 9 8

ISBN-13: 978-0-07-147943-1
ISBN-10: 0-07-147943-0

*The sponsoring editor for this book was Kenneth P. McCombs and the
production supervisor was Richard C. Ruzycka. It was set in Fairfield by
International Typesetting and Composition. The art director for the cover
was Anthony Landi.*

Printed and bound by RR Donnelley.

This book is printed on acid-free paper.

McGraw-Hill books are available at special quantity discounts to use as
premiums and sales promotions, or for use in corporate training programs.
For more information, please write to the Director of Special Sales,
McGraw-Hill Professional, Two Penn Plaza, New York, NY 10121-2298.
Or contact your local bookstore.

To my parents, Prem Vati and Kishan Lal Gupta,
Bill Smith, the inventor of Six Sigma,
and the astronauts of space shuttle Columbia,
for their ethics, vision, and sacrifice, respectively.

CONTENTS

CHAPTER FIVE. PLANNING FOR THE SIX SIGMA BUSINESS SCORECARD   95

CHAPTER SIX. SIX SIGMA BUSINESS SCORECARD DEVELOPMENT   115

CHAPTER ELEVEN. MONITORING PERFORMANCE USING THE SIX SIGMA BUSINESS SCORECARD   197

CHAPTER TWELVE. PERFORMANCE, PROFITABILITY, AND STANDARDS   211

CHAPTER THIRTEEN. LEADERSHIP FOR PERFORMANCE   229

BOB GALVIN'S REFLECTIONS

Because the Six Sigma Business Scorecard takes the Six Sigma methodology to a new level by creating performance standards, it must incorporate input from Bob Galvin, Chairman Emeritus of Motorola, Inc. Bob is the first CEO who has implemented Six Sigma companywide. During my association with Motorola over about 20 years I have looked up to Bob and his team of executives for lessons in leadership.

In 1988 the Leadership section of the Malcolm Baldrige National Quality Award (MBNQA) guidelines was based on Bob's leadership practices when he was CEO of Motorola. His exemplary performance as a leader is still recognized and sought after worldwide. He has a reputation as a visionary. Although he's retired and spends time learning to play the cello and maintaining his energy through athletics, Bob continues to provide leadership advice on various aspects of business and to various sectors, including the U.S. government. Bob is involved in several projects to keep his ever-learning mind active in providing value for society and guidance to Motorola.

I sought Bob's thoughts about the Six Sigma Business Scorecard and corporate performance in the current business environment. His personal vitality, the depth of his convictions, and his enthusiasm for sharing ideas are evident in his reflections.

On Achieving and Sustaining Superior Corporate Performance

Leadership must provide direction in bits and pieces in such a way that it stimulates activities. Leaders must often engage in anecdotal, relevant, and contributory activities. For example, Motorola developed a team competition (TC) that was adopted from Miliken & Company. It created the same type of mentality that permeates a football team competing in the Super Bowl. The TC provided an opportunity for teams to get together and showcase their winning attitudes and successes. These events are large sessions where people really get excited about sharing their work and are enthusiastically led by their CEO.

Such sportslike competitions provide reward and personal stimulation and encourage employees to achieve superior performance. Leaders cheering for their teams demonstrate their commitment to competitive and superior results throughout the corporation.

On the CEO's Role in Achieving the Desired Corporate Performance

Achieving superior performance is personal. The CEO and other high-level executives must commit themselves to improving their personal quality and the people they influence. For example, a phone company executive who attended a Motorola Briefing (to share the experience of winning the MBNQA) decided to improve his personal quality and performance. He enhanced the way he did his job by speaking more nicely to people and becoming punctual for meetings. As a result of these personal changes, in one of his next meetings, meeting attendees all got together and finished the meeting one minute before it was supposed to start!

Quality is translatable. Whether it is a vice president, a manager, a crew supervisor, or a supervisor, personal vitality must be demonstrated in a way that appeals to the employees. Therefore, once the improvement initiative is in motion, it is unstoppable because of the vital spirit of everyone involved. For example, Bill Smith at Motorola, who taught me the concept of

Six Sigma, demonstrated this principle well in how he communicated Six Sigma. I would invite people and ask what they were doing only to discover that a lot of neat things for process improvement were occurring that I could learn (from employees such as Bill Smith).

As for the skeptics, we do not need to do anything with them. They discover for themselves that if they are left out, they will no longer be vital members of the team. At the personal level, these concepts are simple to implement. We must teach employees how to map jobs, determine the time needed to do them, identify opportunities for improvement, and apply statistical thinking and simple tools that have been embraced by people like you and me.

As for the link between quality and profitability, I firmly believe that the link is evident. It must be absolutely clear that when we make fewer mistakes and prevent latent defects from being shipped to customers, the savings are directly transferable. The improvement must lead to lower costs, faster delivery, and higher customer satisfaction. In the Motorola corporate auditing department, when the process was mapped and understood better, auditors realized that the audits could be done in 10 to 20 percent of the time they previously took.

Regarding CEOs or executives who do not see the direct link between poor quality and profitability, I say they must be blind. They may need a little arm twisting. If they do not work out, they must leave the company. The disposition process must be clearly communicated. The boss must evangelize Six Sigma or the improvement process; otherwise, it will not be a successful initiative.

CEOs can communicate their message about quality through case histories, success stories, and anecdotes about experts. Jack Germain, Bill Smith, and Dick Buetow, all leaders in quality, were credible people based on their experience and contributions in their respective professional areas. They were believable people.

I never expected any of my people to be perfect. I wanted them to at least listen to what the leadership was saying and participate in the process. People can participate in the process by simply mapping the way they do their work and discovering opportunities for breakthrough. For example, people in the patent office were doing a fine job. However, as they participated in the Six Sigma process, they discovered that the time used to file patents could be reduced significantly.

Many quality gurus have already taught us that quality never costs money. However, quality improvement, or directly linking to profitability, may need awakening. I hope that *Six Sigma Business Scorecard* will awaken lots of people and be able to change them with help from several "Bill Smiths."

Art Sundry led that awakening at Motorola when he stood up at a management meeting and proclaimed, "Our quality stinks!" People like Sundry have credentials. He was a successful leader who still desired to improve. He kept looking for opportunities to improve and places to find some tools to help him improve. Improvement in performance is an aggregation of a lot of little details.

Different people take different stimuli. At Motorola, we happened to be lucky that we made a point of having an open dialogue in officers' meetings. At such meetings the chief must listen instead of talk. People begin to respond when the leadership listens. We do not change over a weekend. Because of the open communication environment, Bill Smith had the courage to present his idea about Six Sigma, and listeners were willing to entertain it. In such an environment, officers know what needs to be done, take risks, and challenge the existing processes. They find a champion, nurture the seeds of success, and multiply that success into many successes.

Bill Smith wanted to see me, and he took the risk to do so. I immediately listened to him, and when I could not understand his hypothesis, I asked him to come again the next morning.

On Sustaining the Corporate Performance

The CEO of a company is like the captain of a team. If a CEO cannot stimulate the team, he or she does not hold high promise for the company. Leaders such as Roger Miliken of Miliken & Company and John Pepper, ex-CEO of Proctor and Gamble, bring in 10 to 15 like-minded individuals who have similar beliefs. The leader becomes a zealot and goes about promoting his or her quality religion throughout the company. The demeanor and conduct of the institution where quality has emerged radiate throughout the entire corporation. As quality improves, financial measures improve. By giving attention to all, people recognize that employees' contributions are a much-appreciated phenomenon. The CEO must listen and react to employees' ideas.

Speaking of attention, Motorola developed an excellent education program through Motorola University. The objective was to get employees into the classroom for communicating corporate beliefs, to learn the new processes, and for encouraging team competition, where good ones share success stories for others to follow. Personal visibility is a human need. At Motorola we even invited suppliers to learn about our improvement process. All said they also wanted to be better suppliers. Institutionalizing corporate beliefs, values, and goals throughout the theater of operations is critical to sustaining long-term performance. When the CEO's belief permeates each and every employee, the results will surprise us, as we experienced at Motorola from 1987 to 1992.

On Creating a Sense of Urgency to Achieve Superior Financial Results

I set my own deadlines for achievement, and my people had to satisfy their objectives. If someone does not understand the objectives, then the leader has to work with the person. Sometimes we make changes in personnel if the objectives and the current employee skill set appear to be divergent.

Financial measures take care of themselves. If we map the process and do the work in the most effective way, we end up having good results. Such an approach brings down the cost, is appreciated by the customers, and is reason enough for repeat business and even more business.

On Executives' Accountability for Improvement

Every month at Motorola, the top 15 to 25 executives met. One was designated to be the main speaker and to present the best idea to improve quality in his or her area. For that meeting, the executive became a teacher, sharing ideas with his or her counterparts. Whatever the "religion" preached at the company, one must accept it or go to another "church."

On Thoughts about a Participative Management Program

We first learned about the participative management process early, in the 1960s. We took the essence of it, refined it in the Motorola way, and implemented it. This process allowed employees and groups to establish stretch goals to generate savings or execute their tasks better. The savings were shared with employees of the group. This process did wonders for us. We created numerous pockets of excellence and success stories. Employees earned a significant percentage of their salary due to the savings they realized for Motorola.

The CEO Quality Award was created to recognize the extraordinary successes of a team or an individual who has significantly improved a process, leading to higher customer satisfaction or less waste of resources. For people who received the CEO Quality Award, it was a very important award, as the standards to achieve this award were very tough.

On Motorola Today

The Six Sigma and process improvement initiatives were attenuated in the mid-1990s due to company growth. However, they are being restored currently. The way they are being applied today is somewhat different. The program is now called *Six Sigma Digital*. People who are advocating it and living it are very enthusiastic about it. There is a personal vitality to the entire program that comes through Chris Galvin, Chairman and CEO, and Mike Zafirovski, President and COO. They have truly personalized it.

Interestingly, the leaders at GE, Honeywell, Citibank, and other companies all recognize that they have learned something from Motorola that helped them bring about their recent successes. Motorola's being recognized is a rewarding experience.

On Thoughts about the Six Sigma Business Scorecard

An important word is *flexibility*. Any scorecard must not present a rigid construct. The scorecard must allow flexibility to excite the human mind. It is much easier to influence people who are involved in the process.

I believe people will read *Six Sigma Business Scorecard*, get excited about something in it, and generate many ideas for performance improvement. For example, using the title of *Chief Growth Officer* instead of *Chief Technology Officer* makes sense. That is a clever idea. It makes sense to bring the business plan mentality to the technology front.

TAKE AWAY

1. Sports-like competitions provide reward and personal stimulation, and encourage employees to achieve superior performance.

2. The CEO and other high-level executives must commit themselves to improving their personal quality and the people they influence. People begin to respond when the leadership listens.

3. The CEO Quality Award was created to recognize the extraordinary successes of a team or an individual who has significantly improved a process, leading to higher customer satisfaction or less waste of resources.

4. Quality is transferable. Once the improvement initiative is in motion, it is unstoppable because of the vital spirit of everyone involved.

5. People can participate in the process by simply mapping the way they do their work and discovering opportunities for breakthrough. If we map the process and do the work in the most effective way, we end up having good results.

6. Linking quality improvement to profitability may need awakening, and the Business Scorecard can awaken lots of people.

7. The scorecard must allow flexibility to excite the human mind.

FOREWORD

An old adage in business says, "If you can't measure it, you can't manage it." Measuring business performance effectively is a critical responsibility for executive management to ensure improvement in operational effectiveness and sustained profitable growth. Over the past two decades, several business improvement methodologies have come into vogue, including Six Sigma. However, there have been limited advances in the approaches to measuring business performance. In the absence of a holistic approach for measuring business performance, the tendency of managers is to focus on short-term performance improvement, often at the cost of long-term business success. The current Business Scorecard approaches are cumbersome and have failed to meet the test of usability, insightfulness, and actionability. It is ironic that we expect our executives to manage performance precisely quarter after quarter by giving them three color charts with scorecards that represent history which nothing can be done about.

Knowing Praveen Gupta from the dotcom days about six years ago, and working in the areas of business performance, I believe that the Six Sigma Business Scorecard is a unique approach to measuring corporate performance. It is a quantum leap over the much-touted Balanced Scorecard approach. The Six Sigma Business Scorecard places a lot of emphasis on implementation of the measurement system, so that it can actually be used to drive business performance. It includes consideration of all players who determine business performance in a corporation, thus ensuring accountability, collaboration, and innovation. As said in the book, the Six Sigma Business Scorecard mandates inspiration by executives, improvement by managers, and innovation by employees.

Tracking *Six Sigma Business Scorecard* from its manuscript to the current state, I am amazed how rapidly it has caught up with the audience globally, as evidenced by its translation in multiple languages and software products. Executives who are dissatisfied with the approach to measuring performance at their corporations now have a genuine alternative that can provide confidence in measurements and assist in predicting financial results.

The first edition of the book represented the foundational work in the area of corporate performance. The second edition provides detailed guidance on the execution of strategies through performance measurement. Additional chapters provide details on measures of innovation, strategies for execution, and experiences from the field.

In the Six Sigma Business Scorecard, Praveen has developed an elegant and comprehensive methodology for evaluating and predicting corporate performance. The integration of Business Performance Index (BPIn), the implementation methodology, and technology can provide a powerful tool for diagnosing and predicting corporate profitability.

I have enjoyed *Six Sigma Business Scorecard*, and I am sure that it will become an invaluable tool for you to measure and improve the performance of your business.

Mohanbir S. Sawhney
McCormick Tribune Professor of Technology
Director, Center for Research
in Technology & Innovation
Kellogg School of Management

FOREWORD

Now more than ever, corporations the world over are scrambling to redefine the processes, techniques, and strategies they need to survive in an age of uncertainty. The sterling stocks of the 1990s, tarnished by world strife, ethical aberrations, miscalculation, and complacency, are in need of a new framework for survival. Praveen Gupta provides just that by merging two powerful performance improvement processes, Six Sigma and the Balanced Scorecard, to create *Six Sigma Business Scorecard.*

I worked with Praveen at Motorola University for several years. His experience there, at other parts of Motorola, and AT&T Bell Laboratories before that, and his current role as a Six Sigma performance consultant make him well qualified to comment on the fluid approaches needed in our constantly changing corporate climates. Both Six Sigma and the Balanced Scorecard have been used historically to manage functions and operations. By blending these processes, Praveen takes their power to the next level, challenging us to redirect our use of these tools to the measurement of the value-added activities of our efforts. Rather than purely statistical processes, Praveen provides a roadmap for improving profitability and involving employees intellectually, by demanding dramatic improvement from organization leaders.

By implementing the tools offered in Praveen's model, the reader should expect more than process improvement and revenue monitoring. The goal here is the rapid improvement in profit and sustainable business growth.

Many winners of national and regional quality awards have gone on to stumble because they focused on process improvement, defect reduction, and cost reduction without aligning those efforts with customer satisfaction, employee development, business growth, and profit improvement. The Six Sigma Business Scorecard places new emphasis on the latter.

I applaud Praveen for leading business outside of the lines by developing this dynamic hybrid. His powerful formula demands that institutional leadership be held accountable for achieving dramatic rates of improvement, while investing in the intellectual capabilities of all employees. I am sure that you will enjoy and learn as much as I did from *The Six Sigma Business Scorecard*. It is an important work that offers key solutions to the challenges of our times.

A. William Wiggenhorn
Chief Learning Officer
CIGNA
Philadelphia, Pennsylvania

PREFACE

I am pleased at the success of the Six Sigma Business Scorecard in the business intelligence arena. I am thankful to all the readers worldwide who have sent their wishes, comments, and compliments for writing the book. In published reviews, readers have asked for more guidance to implement the Six Sigma Business Scorecard. One of their concerns has been to overcome their current ineffective scorecard due to internal organizational priorities. To help readers of the book, we have developed a simple, yet powerful, Strategy for Execution (SFE) map. Besides, we have also given a case study to demonstrate functioning of the Six Sigma Business Scorecard, even in a small company.

Another group of readers has been asking for measures of innovation. For them, we have provided additional resources for using measures of innovation built in the original Six Sigma Business Scorecard. Finally, we have provided Take Away at the end of each chapter for the ease of understanding.

In the first edition of *Six Sigma Business Scorecard*, I utilized Deming's PDCA (plan, do, check, and act) model to analyze a process. In this edition, I have included a new model of process management, called the 4P model, which focuses on achieving virtual perfection, or excellence, instead of an acceptable performance.

With the addition of the above three chapters, Take Away, and the 4P model, I believe the book has become stronger in guiding its readers to improve business performances.

I would like to acknowledge, not in any order, Arvin Sri of Accelper, Steve DuBrow of i-Nexus, Bill Cooper of Lear Corporation, John Hovis of GoalCentrix, Gary Swanson of The Americas, Juergen Kuebler of Stratascope, Edward Glickman of PREIT, Lonnie Rogers and Barb Schultz of Ideal Aerosmith, Paul Davis of Scanlon Leadership, Anoop Verma of CNA, T.J. Teck of Access Electronics, Cissy Pettenon of PSA, Darsh Wasan and Bob Anderson of IIT Chicago, Venkat Ravilla and Rajiv Khanna of Gantec USA, Prof. Scott Young of Department of Management at DePaul University, Brian O'Reilly of Dentsply, Baber Inayat and Glen Nevogt of Superior Essex, Phil Armour of Corvus International, Rick O'Brien of Payment Pathways, Tom Crowley of MBX, and Anton Hirsch of Cobra Metal Works, for their support to *Six Sigma Business Scorecard* or my work in this area. My sincere thanks to Rajesh Tyagi of DePaul University for recognizing the potential of the Six Sigma Business Scorecard, and collaborating to develop it further. I would like to express my gratitude to Prof. Mohan Sawhney, Kellogg School of Management, for staying involved for the last several years, and writing Foreword for this edition of the book. Finally, I thank Ken McCombs and the staff at McGraw-Hill for their unwavering support to the book.

I look forward to hearing from you, the reader, again about your experience with the Business Scorecard or the book.

Praveen Gupta
Accelper Consulting
praveen@accelper.com

ACKNOWLEDGMENTS

I am grateful to my family (Archana, Krishna, and Avanti) for allowing me to write this book at odd hours, and to many friends and colleagues for listening to my unsolicited and painful ideas. My sincere thanks to Frank Brletich and Caryn Pernu for their review, recommendations for, and edit of the manuscript. Thanks to Rajiv Varshney for jazzing up the figures and Rajeev Jain, Jim McNulty, Barb Schultz, and Terry Luczak for their thoughtful discussions.

The book would not have passed beyond the thought process without knowing Ken McCombs, Senior Editor at McGraw-Hill. Ken has been phenomenal in believing in the Six Sigma Business Scorecard concept and in helping patiently during the development of this book. The staff at McGraw-Hill has done a wonderful job of making *Six Sigma Business Scorecard* a reality.

I am thankful to Bob Galvin, Chairman Emeritus, Motorola, Inc., for sharing his views with readers of this book, and Bill Wiggenhorn, Chief Learning Officer, Cigna, for reviewing the manuscript and preparing the Foreword. Bob Galvin was the first CEO to launch Six Sigma at Motorola, and Bill Wiggenhorn was the catalyst for institutionalizing Six Sigma at Motorola. I am honored to receive encouragement from these two great leaders.

INTRODUCTION

Businesses today seek an effective corporate performance measurement system to maximize the bottom line. With the advent of the Internet, ongoing globalization, and standardization in management systems, business leaders must focus on how to measure performance to monitor their continued viability and success. Many existing performance measurement systems were designed to support business practices and to monitor progress. With shrinking margins and competitive pressures, however, corporate performance measurement systems must do more than monitor. They must identify opportunities for optimizing profitability and growth, without pitting one against the other. The idea is to use performance measures to add value, instead of simply measuring for a formality. That is precisely what the Six Sigma Business Scorecard is designed to do.

Six Sigma Business Scorecard combines the Six Sigma methodology for dramatically improving customers' delight with the Balanced Scorecard method for achieving financial objectives. The Six Sigma Business Scorecard utilizes strategy to maximize profitability and growth, accelerate improvement, foster leadership accountability, and encourage employee involvement. It builds on proven methodologies and offers a new method for measuring corporate performance. Filling in the blanks that have been left open by the other measurement systems or methodologies, the Six Sigma Business Scorecard is integrated into a Business Performance Index (BPIn), a measure of corporate wellness. Currently, there is no method that enables an organization to determine its overall performance level. The Six Sigma Business Scorecard methodology highlights the responsibilities for corporate performance and establishes a formula for determining the sigma level of an organization.

The purpose of the Six Sigma Business Scorecard is twofold: (1) to identify measurements that relate key process measures to a company's profitability, making the opportunities so visible that they are difficult to ignore, and (2) to accelerate the improvement in business performance. Optimizing the profitability, cost, and revenue variables is a primary purpose of the Six Sigma Business Scorecard.

The salient features of the Six Sigma Scorecard include the following:

- Provides a new model for defining a corporate sigma level
- Aligns with the business's organizational structure
- Maintains visibility of cost, revenue, and profitability
- Includes leadership accountability and rate of improvement

The Six Sigma Business Scorecard is a great model for establishing a common target in terms of corporate *defects per unit* (DPU), *defects per million opportunities* (DPMO) and Sigma. In the absence of such a target, corporate leaders are unable to focus and suffer from ineffective implementation of Six Sigma. The Business Performance Index provides a one-number measure of corporate performance in terms of Sigma that can be used as a benchmark for driving future improvements. The benefits of the Six Sigma Business Scorecard include the following:

- Provides a target for performance improvement
- Enables a business to drive dramatic improvement
- Promotes the intellectual participation of all employees
- Forces changes in an organization on a continual basis

- Acts as a catalyst for bringing out the best among employees
- Generates energy and enthusiasm
- Reduces cost and improves profits

Throughout the book, a business has been considered as a collection of business processes, irrespective of their output, products, or services. Therefore, the Six Sigma Business Scorecard can be applied to all organizations. It has been adapted for even small businesses based on the feedback from small business owners. The book is meant to be a documented methodology to implement a practical corporate performance measurement system that provides some predictable indicators and identifies opportunities for improvement. To benefit from the Six Sigma Business Scorecard, business leaders must be willing to think creatively and make necessary organizational changes to balance profitability and growth. The organization of the revised *Six Sigma Business Scorecard* is shown in Figure I-1.

Six Sigma Business Scorecard includes 16 chapters that lead step by step through the implementation process. Different organizations will face different challenges in implementing the Six Sigma Business Scorecard, depending upon their complexity and level of commitment to improving profitability and growth. Many in the leadership cadre believe that profitability and growth have an adversarial relationship. The current business environment encourages maximization of one or the other. However, to survive and thrive, businesses must balance both profitability and growth.

The Business Performance Index has been validated based on estimation and public information about the companies on the Dow Jones Industrial Average Index and several companies on a one-to-one basis. BPIn has shown correlations with performance. BPIn would not be significantly affected by minor errors in the measurement system. To improve a company's BPIn, that is, profitability and growth, the company must improve dramatically in all areas of business led by the leadership, facilitated by management, and executed by employees.

FIGURE I-1. Layout of *Six Sigma Business Scorecard.*

With the renewed focus on leadership for corporate performance, the Six Sigma Business Scorecard can be used as a tool that will allow an organization's leadership to keep profitability in sight, involve employees intellectually, and demand dramatic improvement from the management team. Achieving profitability and growth requires a unique recipe for each business. The Six Sigma Business Scorecard allows that recipe to evolve for the benefit of all stakeholders. For organizations to succeed, the leadership must inspire, managers must improve, and employees must innovate. This combination of roles will produce strong teamwork, corporate synergy, and excitement and resonating results that are visible and felt by everyone in the organization.

Having been involved with the Six Sigma methodology from its early days, I believe no corporation monitors Sigma level or overall performance index. With the Six Sigma Business Scorecard and BPIn, businesses can now determine the corporate wellness index and Sigma level that should be a leading indicator of the financial performance.

Given that *Six Sigma Business Scorecard* offers a new methodology, there will be opportunities for improvement and further refinement in the methodology. The author plans to monitor companies that successfully implement the Six Sigma Business Scorecard and those that struggle against implementing it. The author would be greatly honored to receive any feedback for further refinement of the Six Sigma Business Scorecard at praveen@accelper.com.

SIX SIGMA
BUSINESS
SCORECARD

TRENDS IN PERFORMANCE MEASUREMENTS

To survive and grow, a business enterprise must meet its stated objectives. How do we know if a business is meeting its goals? Just as we go to a doctor for a checkup to confirm our physical well-being, a business also needs monitor its health. It needs to measure its performance frequently to evaluate its past performance and identify areas where it must improve in the future.

Historically, as supply geared up to meet the rising demand for consumer products, the productivity of an enterprise became the most important performance measure. As the gap between supply and demand decreased and competition increased, customers had better choices, and they began demanding more features and better service. Customer demand propelled companies to innovate and diversify in order to grow, and businesses became more complex. Productivity alone thus became an insufficient measure of business performance.

TECHNOLOGY AND GLOBALIZATION

The advent of transistors was the next big innovation in business. Transistors accelerated the growth of computing and information technology, leading to applications that helped produce products and services that were cheaper, more accurate, and of higher quality. A 1998 automobile print advertisement boasting that its latest cars possess more electronics and computing power than did the Apollo 13 is an example of how striving for increases in product functionality has led to more technological advancements. These technology developments fundamentally changed the way business is done.

With technology, distances have grown shorter, and the world has become one large market. The exponential advancements in telecommunications and Web-based technology reduced the virtual travel time between two places to mere seconds. In other words, information can be exchanged almost anywhere around the world within seconds. These advancements helped to reduce the effect of time zone differences and cultural barriers, and distributed production worldwide. A global company in the apparel market with retail stores in the United States may have its products produced in factories in China, may manage its business with Enterprise Resource Planning software coded in Israel, and may have its customer call center staffed by representatives in India.

The global integration and increasing competition also created challenging business dynamics. For example, the electronics industry's explosive growth and competitive pressures resulted in reduced margins despite high innovation and excellence in performance. The true challenge is to achieve growth through innovation and excellence while maintaining high profitability.

In addition to changes brought about by technology and globalization, businesses face an ever-evolving landscape of frequent mergers and acquisitions, competition from parallel and substitute products, and threats from buyers and suppliers. Excelling in this environment requires a comprehensive reporting system that can accurately read the market's pulse. A performance measurement system (e.g., a report card or scorecard) that provides feedback for timely response, as well as monitoring and enhancement purposes, must exist.

MACRO ECONOMIC MEASURES

The U.S. economy is a classic example of a feedback system constantly monitored through a set of measures (see Figure 1-1).

Year	GDP	PPI	CPI	PI	VPI	CU	TIPI	E/H ($)	O/H Index	GI	Profits
	($T)			(000)		(%)				($B)	($B)
1970	1.04	39.3	38.8	4.1	56.3	81.1	58.7	3.23	67	152.4	28.5
1980	2.8	88	82.4	10.2	30.2	81.5	79.7	6.66	80.4	477.9	92.5
1990	5.8	119.2	130	19.6	40.6	82.3	98.9	10.01	95.4	861.7	110.1
2000	9.9	138	172.2	30.2	47.9	82.1	147.5	13.75	101.4	1767.5	275.6

GDP: Gross Domestic Product   CU: Capacity Utilization
PPI: Producer Price Index   TIPI: Total Industrial Production Index
CPI: Consumer Price Index   E/H: Earnings per Hour
PI: Personal Income   O/H: Output per Hour
VPI: Vendor Performance Index   GI: Gross Investment

FIGURE 1-1.   Measures of U.S. economic performance, 1970–2000. (*Strawser, 2001.*)

These national economic indicators are signals that investors, business, and government respond to in order to encourage long-term growth of the U.S. economy. The U.S. federal government, for example, reacted with tax and interest rate cuts in 2002 to spur the economy after measures such as the Gross Domestic Product (GDP) growth rate, unemployment, and capital spending reflected a slowing economy. The collection of data, analysis, and corrective action must occur a short time after the actual event to counter undesired outcomes.

EVOLUTION IN PERFORMANCE MEASUREMENTS

After World War II, several national economies grew significantly, leading to a global competitive environment. From time-motion studies to quality improvement tools, businesses deployed methods to improve their performance. Beginning in the 1970s, Japanese automakers challenged U.S. industry by deploying quality management tools taught by J. M. Juran, Edwards Deming, Phil Crosby, Genichi Taguchi, and others. In the 1980s, other ways to promote process and performance standards were created, such as the ISO 9000 quality management system developed by the International Organization for Standardization (ISO) and the Malcolm Baldrige National Quality Award (MBNQA) guidelines established by the U.S. Motorola-pioneered Six Sigma methodology and successfully implemented to reap rich benefits. Figure 1-2 shows the evolution of various techniques.

The purpose of these new quality management techniques was to improve the performance of business processes. The process control charts developed by quality pioneer Dr. Walter Shewhart brought together the disciplines of statistics, engineering, and economics to improve the consistency of production processes. Design for Manufacturability helped improve reproducibility, recognizing the importance of considering cost, quality, and manufacturing characteristics early in a product's design stage. The ISO 9000 system improved interrelationships between business functions. Benchmarking helped in competitive positioning by measuring comparative operating performance and identifying best practices. Six Sigma accelerated the rate of improvement. Lean manufacturing increased process agility, and innovation introduced new products faster. The improvements in business performance also led to an increase in customer expectations and demand for products

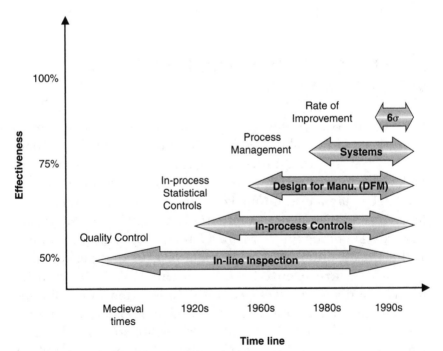

FIGURE 1-2.   Performance control methods.

that were better, faster, and cheaper. The spiral of demand and supply changed the paradigm from a focus on quantity to a focus on quality and functionality.

To meet customer expectations for better value and shareholder expectations for bigger profits, a new system for performance measurement and improvement must evolve. From financial to process measures, a balanced approach is required—an approach that encompasses strategic, operational, and leadership functions.

JURAN'S FINANCIAL AND QUALITY TRILOGIES

J. M. Juran developed the Financial and Quality Trilogies (see Figure 1-3), which have three aspects: planning, control, and improvement. The Financial Trilogy elements of planning, control, and improvement all end up at sales—planning for sales, controlling sales, and increasing sales—leading to a strategic sales plan. The sales plan includes sales by segments, budgets allocated to achieve those sales, margin goals, resource requirements, distribution requirements, and discount models. The CEO or equivalent typically develops this strategic sales plan, which is reviewed quarterly to see how well it is being executed and to develop follow-up action items.

The Financial Trilogy works well in a supply-driven environment, but it has limitations. When results are not achieved,

FIGURE 1-3.   Financial and QualityTrilogies. (*Juran, 1998.*)

the model offers no system to change the course (except to change the executives or the salespeople). Furthermore, the Financial Trilogy does not include all employees, relate to operational performance, or drive internal improvement. In addition, when the market forces change in the presence of ever-increasing competition, such a model creates chaos, stress, and poor performance because it leaves many workers out of the loop rather than encouraging the intellectual involvement of all workers across departmental boundaries.

The Quality Trilogy elements of planning, control, and improvement involve strategic quality planning, process controls, and product improvement. The Quality Trilogy is typically the responsibility of the quality manager or equivalent, who plans for quality through data analysis; controls the quality through inspection, test, and repair; and improves product quality through programs such as Total Quality Management (TQM). All aspects of the Quality Trilogy are managed by the quality department in a subservient relationship to management. Because the quality manager is held responsible for defective products, one challenge with the quality trilogy is that it points blame to the quality department for production problems. In other words, it lacks accountability and leadership. In addition, history has shown that conventional quality programs do not show a significant impact on profitability; hence they often look like investments without meaningful results.

BUSINESS TRILOGY

In the 1970s and 1980s, we saw a tremendous shift in the role of the quality department with the advent of ISO 9000, the Malcolm Baldrige National Quality Award, and Six Sigma. Customer expectations about product quality changed from simply looking for a good warranty program to completely rejecting (with penalty) any supplier who could not provide quality products. Suppliers were held responsible for damage caused by their poor performance. Customer expectations for feature-rich and innovative products, extraordinary customer service, and the lowest available price brought about the Business Trilogy model. In the Business Trilogy (as shown in Fig.1.3), executives plan based on input from employees, the management team exercises control over processes, and improvement occurs when all minds in the company work toward it. But leadership more than management is what's needed to create a superior company. Leadership inspires employees toward a common goal, defining their purpose, communicating the consequences of low performance, and rewarding excellence rather than mediocrity. We've all seen executives who created several multimillion-dollar problems and were rewarded because they then solved those problems! Examples abound of leadership teams that implement poorly planned, controlled, and executed decisions, yet who still receive incentives in hopes of better results next time. Without new systems and processes, the second time around is not going to be any better than the first.

Any business has three segments: leadership, operations, and support. The business theater of operations must include superior leadership, excellent management, and committed employees. In such a model, the leadership plans to become a superior company through proper goal setting for higher profit margins. The management team manages processes for dramatic improvement (highest quality in the shortest time). Employees produce innovative products or deliver services with love and care. Of course, what goes around comes around. Employees will only care when they are cared for by the leadership.

ISO 9000 QUALITY MANAGEMENT SYSTEM

The ISO 9000 system evolved when the European Union (EU) was forming in the mid-1980s. Its main purpose was to provide standards that would facilitate trade between EU member countries. Later it became an international standard for quality management under the auspices of the International Organization for Standardization (ISO). In reality, the ISO 9000 system is a business management system. As products and services became complex and customer requirements became stringent, the responsibility for business performance moved from the quality department to all departments. Today, quality or goodness of process and product is the overriding responsibility of everyone in any organization.

The newest version of ISO 9000 includes additional measurements that weren't in the original plan, as well as the requirement for continual improvement. It has the following requirements:

4.0   Quality Management System

- General requirements
- *Documentation requirements.*   General, quality manual, control of documents, control of records

5.0   Management Responsibility

- Management commitment
- Customer focus
- Quality policy
- *Planning.*   Quality objectives, quality management system planning
- *Responsibility, authority, and communication.*   Responsibility and authority, management representative, internal communication
- *Management review.*   General, review input, review output

6.0   Resource Management

- General
- Provision of resources
- *Human resources.*   General, competence, awareness and training, infrastructure, work environment

7.0   Product Realization

- Planning of product realization
- *Customer-related processes.*   Determination of requirements related to the product, review of requirements related to the product, customer communication

- *Design and development.*   Planning, inputs, outputs, review, verification, validation, control of changes
- *Purchasing.*   Purchasing process, purchasing information, verification of purchased product
- *Production and service provision.*   Control of production and service provision, validation of processes for production and service operations, identification and traceability, customer property, preservation of product
- Control of monitoring and measuring devices

8.0   Measurement, Analysis, and Improvement

- General
- *Monitoring and measurement.*   Customer satisfaction, internal audits, monitoring and measurement of processes, monitoring and measurement of product
- Control of nonconforming product
- Analysis of data
- *Improvement.*   Continual improvement, corrective action, preventive action

As the list illustrates, the requirements are quite comprehensive. The ISO 9001:2000 system is based on process thinking:

1. Plan for setup.
2. Do for excellence.
3. Check for verification.
4. Act for improvement.

PROCESS THINKING

To perform well, the process owner needs to control inputs, process activities, and the process output. *Control* means that the process owner understands the requirements as well as receiving, producing, or supplying according to the requirements. Verification methods must be implemented at various stages of the process to ensure that requirements are being met. This might be done by controlling input material through supplier qualification, as well as monitoring the process by means of data analysis, inspection, tests, and measurements. A corrective action is initiated if the verification shows that the requirements are not being met.

A business benefits from consistency of operations once the business infrastructure is established using the ISO 9001 quality management system. However, even with an ISO 9001 system, the business will operate at below desired performance levels if the processes are not optimized.

MALCOLM BALDRIGE NATIONAL QUALITY AWARD GUIDELINES

When the ISO 9000 quality management system was launched in the United States in 1987, the U.S. Congress established the Malcolm Baldrige National Quality Award to improve business performance. The resulting MBNQA Criteria for Performance Excellence (also known as the Baldrige Criteria) provided a system for managing process performance. The first Baldrige award went to Motorola, where I had an opportunity to contribute to the quality systems while working with the late Bill Smith. The purpose of the Baldrige Criteria was to accelerate the rate of improvement and innovation. The award criteria provided a performance framework based on measures of leadership, customer information, operations, human resources, and finances. Studies showed that organizations implementing systems based on the award guidelines outperformed their counterparts in value creation, stock performance, and market leadership.

With worldwide acceptance, the Baldrige Criteria raised awareness of customer satisfaction, leadership, and information management. The Baldrige Criteria were based on global best practices. For example, the leadership category was created based on behaviors and practices deployed by leaders such as Bob Galvin, then CEO of Motorola. Leadership included definition of a vision, communication with employees, community service, public responsibility beyond the organization, and constancy of purpose. The Baldrige Criteria provide a role model for companies to achieve superior performance, while ISO 9000 provides a framework for an organization to achieve good results consistently.

The Baldrige Criteria highlight the following aspects of business performance:

- Visionary leadership
- Customer focus
- Employee development
- Process excellence
- Market leadership
- Superior financial results

These business aspects are addressed in the seven categories of Baldrige Criteria for excellence, as shown in Figure 1-4, as

Category	Area	Point Value
Leadership	Organization Leadership Public Responsibility and Citizenship	120
Strategic Planning	Strategy Development Strategy Deployment	85
Customer and Market Focus	Customer and Market Knowledge Customer Relations and Satisfaction	85
Information and Analysis	Measurement and Analysis of Organizational Performance Information Management	90
Human Resource Focus	Work Systems Employee Education, Training, and Development Employee Well-Being and Satisfaction	85
Process Management	Product and Service Processes Business Processes Support Processes	85
Business Results	Customer-Focused Results Financial and Market Results Human Resource Results Organizational Effectiveness Results	450

FIGURE 1-4. Baldrige Criteria for performance excellence. (*MBNQA Guidelines, 2003.*)

well as in several areas within each category. Each area consists of a set of questions that focus on certain aspects of a business. The objective is to emulate, as much as possible, each guideline in order to achieve superior results.

The three major performance measurement and improvement systems of ISO, the Baldrige Criteria, and Six Sigma all launched at about the same time. ISO launched ISO 9000 and Motorola launched the Six Sigma methodology in 1987, and Congress launched the Baldrige Criteria in 1988. Figure 1-5 highlights key differences among the three approaches.

Faced with fierce foreign competition in the semiconductor industry, the late Bill Smith invented the Six Sigma methodology in 1985–1986. He realized that the rate of improvement Motorola required to be profitable was so significant that something dramatically different had to be done. Motorola

ISO 9000	MBNQA	Six Sigma
A framework for creating "Quality Thinking."	A framework for creating "Performance Thinking."	A framework for linking improvement to profitability.
Facilitates process management through documentation and compliance.	Facilitates benchmarking to improve performance levels to best-in-class levels.	Facilitates dramatic improvement to achieve performance excellence.
Specifies all business functions except Accounting.	Specifies key aspects of business.	Specifies a methodology for improvement irrespective of functionality.
Promotes Management Responsibility through communication and management review.	Promotes exceptional leadership behaviors as a way of life in society.	Requires leadership to aim at highest performance with highest profitability.
Main aspect is compliance to documented practices and improving effectiveness.	Main aspect is to achieve total customer satisfaction through superior practices and performance.	Main aspect is achieving and maintaining a high improvement rate for business aspects that affect profitability.
About 500,000 companies have implemented it worldwide.	About 4 to 8 companies win the national level; similar number at state level and in other countries.	Has been adopted by several companies to achieve dramatic improvement and profitability.
Savings are difficult to quantify.	Performance of publicly traded companies has shown advantage over the others by 3 to 4 times.	Companies have reported huge amount of savings in production and service areas.
Mass application of the standards.	Limited to a few companies.	Selectively used by companies committing to be a superior company.
It is a third-party certification.	It is recognition for excellence.	It is a methodology to optimize performance and maximize profitability.
Is on decline due to diversification in series of industry-specific standards.	Stabilized due to limited recognition. Has expanded into health care and education.	Growing rapidly as an attractive means to realize superior financial results.

FIGURE 1-5. Comparison of ISO 9001:2000, MBNQA, and Six Sigma system requirements.

leaders envisioned the state of manufacturing 20 years into the future. They predicted that cellular phones were going to follow a similar route as the commercialization of digital watches. Interestingly, that is exactly what happened. Today, the trend in the cost of a cellular phone is almost like that of a digital watch several years ago. Like the price of digital watches, the price of cell phones has dropped precipitously. Sometimes a cellular phone is even complimentary with the purchase of a cellular service. This is why new processes for improvement must be created continually and not just as a one-time deal. Sustaining the Six Sigma process is critical to maintain performance once the initial success is achieved.

Before launching the Six Sigma initiative, Bill Smith had discovered that most customer complaints and failures of the product in the field were primarily caused by errors that the quality control system was unable to detect before the product was shipped. Typically, management focuses on resolving customer complaints through developing a superior customer service team. With Six Sigma, however, improving customer service for fixing field failures may be less important than developing systems that will catch and correct errors before the customer receives the product. The field failures really determine the customers' perception of an organization's products or services. To address the customers' concerns, therefore, internal operational excellence, driven passionately by the leadership, is mandatory.

In addition to widespread programs such as ISO 9000, the Baldrige Criteria, and Six Sigma, several companies have developed their own ways of achieving excellence. These can be effective, but issues arise in the methods such organizations use to rate their performance. Every organization looks at financial performance critically every month or quarter. However, many organizations, small or large, do not review their operational performance as rigorously as they do their financial performance. In essence, they are watching the pennies at the end of the line while ignoring the dollars in the line, which should be subject to even greater scrutiny. Companies with superior performance do so and reap the benefits. However, most businesses struggle to find a method of looking at these costs that is simple, practical, and compatible with profitability.

BALANCED SCORECARD

Robert Kaplan and David Norton introduced the Balanced Scorecard in 1991 in *The Harvard Business Review*. The Balanced Scorecard was a relief for many organizations; it gave management a new way of monitoring a company's performance by measuring past success and setting goals for the future. The Balanced Scorecard is designed to provide a strategic vision for the organization by looking at four perspectives: financial, customer, learning and growth, and internal business processes. For each of these four areas, the company looks at goals (How will we define success?), measures (What will we measure to know we are successful?), targets (What quantitative results do we wish to achieve?), and initiatives (What activities will we do to achieve them?).

The Balanced Scorecard has been used by senior executives for more than a decade as an excellent tool for strategy development. Kaplan and Norton demonstrated how executives in industries such as banking, oil, insurance, and retailing have used the Balanced Scorecard to guide their current performance and target future performance goals. However, implementation at the grassroots level has been questioned for practicality.

SIX SIGMA BUSINESS SCORECARD

The business environment has changed a lot during the last decade. Several hundreds of thousands of businesses worldwide implemented ISO 9000 quality management systems. About a million copies of Baldrige Criteria are shipped to businesses annually. Many businesses and their suppliers have implemented the Six Sigma methodology. However, the challenge is to improve profitability significantly, to prevent marketplace dynamics such as the dot-com meltdown of 2000. The value of performance measures has become more important than ever.

With the current and anticipated unsettling business environment—accelerating trends in technology, high expectations for performance, eroding prices, and shrinking profitability margins—businesses need a performance measure (or score-card) that is robust and that addresses various aspects of a business including the marketplace dynamic. Businesses need a tool that provides a framework and guidance, creates challenges, and stimulates excitement. Businesses need performance measures that continually renew and reenergize them, forcing them to discard the status quo and embrace innovation on a continual basis.

Such a system must rely on the basics of a business. A business is a collection of processes, including the leadership process. Each business process has inputs that include suppliers, assets, resources (capital, material, people), and information. The process also has a vision, measures, policies and procedures, and output that includes products or services for customers. Every business has variances. The question is what to do with them. Each business must have a process to handle excessive variances in the organization. Typically, these variances are the leaks in profitability. To fine-tune profitability, one must look at measures of all aspects of the organization the way it really works in order to reverse any loss of profitability.

The Six Sigma Business Scorecard has been developed to look at measures of all aspects of the organization. It addresses the concerns that executives express about current scorecard systems, such as their ineffectiveness at relating to the employees who do the work. Most scorecards are strategic in intent and do not flow down to process measures.

For any scorecard to be implemented successfully, the scorecard must include sound planning, operational excellence, and sustainable growth. With an understanding of the Business Trilogy, process model, and dynamic economic environment, a Six Sigma Business Scorecard was developed that personifies leadership and management; aligns purchases and operations; drives customer service and sales; and promotes employee excellence, innovation, and improvement. Such a scorecard should intuitively be persuasive to executives for strategy as well as rewarding to employees for continual excellence through innovation.

The Six Sigma Business Scorecard, as shown in Figure 1-6, is driven by those responsible for inspiration, planning, and

IMPROVEMENT

INNOVATION

FIGURE 1-6.   Six Sigma Business Scorecard.

profitability (i.e., the leadership); controlled by managers who improve processes and reduce costs; improved by employees who develop innovative solutions to meet customer needs; and steered by sales and customer service representatives who acquire and maintain customers through high-quality relationships for revenue and growth.

TAKE AWAY

1. Customers demand propel companies to innovate and diversify to grow resulting more complex businesses. Productivity alone thus becomes an insufficient measure of business performance.

2. In addition to changes brought about by technology and globalization, businesses face an ever-evolving landscape of

frequent mergers and acquisitions, competition from parallel or substitute products, and threats from buyers and suppliers.

3. The collection of data, analysis, and corrective action must occur in a short time after the actual event to counter undesired outcomes.

4. To meet customer expectations for better value and shareholder expectations for bigger profits, a new system for performance measurement and improvement must evolve. An approach must encompass strategic, operational, and leadership functions.

5. The business theater of operations must include superior leadership, excellent management, and committed employees.

6. To perform well, the process owner needs to control inputs, process activities, and the process output.

7. Every organization looks at financial performance critically every month or quarter. They are watching the pennies at the end of the month while ignoring the dollars in operations.

8. The Balanced Scorecard gave a new way of monitoring a company's performance by measuring past success and setting goals for the future. However, implementation at the grass root level has been questionable at the best.

9. With the current and anticipated unsettling business environment—accelerating trends in technology, high expectations for performance, eroding processes, and shrinking profitability margins—businesses need a performance measure that is robust and that addresses various aspects of business including the dynamic marketplace.

SIX SIGMA—AN OVERVIEW

In the early 1980s, the U.S. electronics industry was in fierce competition with offshore manufacturers. Because of the oil crisis of the late 1970s, the economy was in a slump. Motorola soon realized it had to do something different. The leaders at Motorola envisioned trends in manufacturing in the communications industry. The tack they decided to take was to develop Six Sigma. The leaders conducted benchmarking studies, analyzed problems, and reviewed customer feedback. They discovered that customers expected better products than they were getting and that problems in the field were extensions of internally observed problems. The leaders determined that the future of manufacturing electronic products looked like the future of digital watches—trinkets that had to be produced at a very low cost in order to be a viable product line. Interestingly enough, cell phones and digital watches look pretty much the same today—in price as well as appearance. Today, however, cell phones are offered as part of a service plan instead of being sold as a separate entity.

COMPLEXITY AND PERFORMANCE

As Motorola's leadership gained knowledge and understood the need, the Six Sigma process started to become apparent in statistical terms. The issue was this: Each process by itself could be fine; however, when a series of processes was involved in the production of a product or service, the net was much less than the yield at each process. For example, if process and parts yields were at 99 percent for each component of the operation, the overall yield for the entire operation would be much less, as shown in Figure 2-1.

The overall yield is calculated by multiplying the process yield at each process. In the case of 100 components, this means 0.99 (for 99 percent yield) is multiplied 100 times. The figure shows that as soon as the process or the product complexity exceeds 100, the probability of making errors increases such that almost every unit of output must be repaired at least once. Therefore, a performance level of 99 percent does not stretch far enough for complex products or services.

COST OF POOR PERFORMANCE

With any failure rate, there is an associated cost. The cost of the product is directly related to the failure rate, as shown in Figure 2-2. The total cost is determined according to the formula

$$\text{Total cost} = \text{Unit cost}\,(1 + \text{Failure rate})$$

The additional cost is a waste of valuable resources that is deducted from the overall profitability of the business. The

Number of Operations or Components	Overall Yield, % (99% Yield at Each Process, or for Each Component)
100	36.7
500	0.5
1000	0
5000	0

FIGURE 2-1. Components versus overall yield.

FIGURE 2-2.   Failure rate versus total cost.

waste occurs in all functions of a business, e.g., sales, design, purchasing, production, quality assurance, and even management. Typical accounting practices do not consider all aspects of the business, as the focus is on the visible waste. However, hidden costs need to be understood, identified, and monitored in order to maximize an organization's profitability.

The cost factors, as shown in Figure 2-3, include the direct cost of material, the cost of resources required to handle it, and even opportunities lost because of customer dissatisfaction. Following is a list of aspects of costs that contribute to waste and harm profitability.

- Awareness of common purpose
- Employees' skills and experience
- Defective materials and excessive use of material
- Insufficient information
- Improperly maintained machines
- Wrong or damaged tools
- Impractical procedures
- Lack of training
- Reduced capacity due to failure rates
- Lack of preventive maintenance

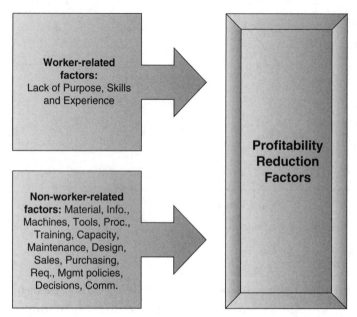

FIGURE 2-3.   Factors contributing to reduction in profitability.

- Suboptimized designs
- Poorly understood customer requirements
- Overambitious sales representatives
- Excessive discounts
- Informally communicated requirements to suppliers
- Purchasing policies requiring cheapest parts instead of value
- Management uncertainty in decision making and lack of direction

Note that most of these cost factors are related to neither management nor employees. However, with changing business practices, increasing complexity of products, and interdependence of various functions, everyone must work to ensure that necessary activities occur correctly. Management is no longer solely responsible for every action of all employees, nor are employees solely responsible for every decision that management makes. Instead, everyone must be responsible for inefficiencies in the system. The system is not a function of two processes; it is a collection of various processes that generate hidden and visible waste in the system.

Considering waste in the system, as well as ever-increasing customer expectations, Motorola recognized that a new yield model was needed. The late Bill Smith, a senior manager at Motorola, developed a model that would allow higher yields at each operation, thus generating better products and services to customers. With the Six Sigma level of performance, each function must be performed virtually perfectly. The failure rate at each operation is expected to be 3.4 parts per million versus simply 99 percent. With such a high level of performance, the field performance improves significantly, waste reduces dramatically, and profitability can improve enormously. Figure 2-4 shows the impact of the revised business model on the overall performance of the system. With this magnitude of improvement, the Six Sigma concept becomes a powerful strategy to drive improvement in an organization's performance and profitability.

BASICS OF SIX SIGMA

Six Sigma offers a measure of goodness, a methodology for improving performance, a measurement system that drives dramatic results, and a new paradigm that requires a passionate commitment from leadership to set high expectations. Figure 2-5 demonstrates the relationship between defect rate, sigma level, and cost reduction opportunities.

Number of Operations or Components	Overall Yield, % (99% Yield at Each Process, or for Each Component)	Overall Yield, % (99.9996% Yield at Each Process, or for Each Component)	Improvement, %
100	36.7	99.9	Greater than 100%
500	0.5	99.8	Greater than 1000%
1000	0	99.6	Practically infinite
5000	0	98	Practically infinite

FIGURE 2-4. Performance improvement due to the Six Sigma process capability.

Defect Rate, parts per million	Sigma Level	Cost Reduction Opportunities, % of Sales
66,810 (or 6.7%)	3	25
6210 (0.6%)	4	15
233	5	5
3.4	6	1

FIGURE 2-5.   Sigma level and related opportunities for improvement.

Companies that have implemented Six Sigma have reported enormous savings. For example, Motorola, GE, Honeywell (formerly Allied Signal), Raytheon, ABB, and many more have realized savings in the hundreds of millions of dollars. During the first 5 years of Six Sigma implementation, Motorola reported about $1 billion in manufacturing operation savings and similar savings in nonmanufacturing operations. For Motorola, 1987 to 1992 was a significant period when sales and profitability improved significantly.

TRADITIONAL APPROACH TO SIX SIGMA

The traditional approach to Six Sigma involves the steps that focus on discovering customers' critical requirements, developing process maps, and establishing key business indicators. After these steps are completed, the business moves on to review its performance against the Six Sigma standards of performance and takes actions to realize virtual performance. The often overlooked aspect of achieving dramatic improvement in business performance is the superior management review process, in which the senior executives become extensively involved in monitoring performance and demanding necessary improvements from their middle managers and employees. Once success is achieved, reward and recognition are critical success factors to perpetuate the rate of improvement.

The traditional approach, used by Motorola during the first 5 years, requires leadership and managerial staff to undergo extensive training for change management supported by Motorola University, followed by training in Six Sigma methods. Senior management then taught employees to create an upward communication process. When employees had a problem, they were required to go to their supervisors for answers. The outcome was a common understanding of the Six Sigma process and common goals for improvement.

Employees were asked to set goals that would allow them to stretch their imaginations, promoting creativity and superior teamwork. When teams excelled, they were recognized by the CEO and were rewarded with the CEO Quality Award. This became the best award employees could receive, and they strived for it.

THE BREAKTHROUGH APPROACH TO SIX SIGMA

The new approach to Six Sigma, called the *Breakthrough approach* and developed by Mikel Harry and Richard Schroeder (2000), captured the Motorola methods and packaged them in the *Define, Measure, Analyze, Improve, and Control* (DMAIC) methodology. The Breakthrough approach consists of management involvement, organizational structure to facilitate the improvement, customer focus, opportunity analysis, extensive training, and reward and recognition for successful problem solving. Its benefits include the standardization of the methods, global adaptation of the methodology, and commercialization of Six Sigma.

In any case, whether Six Sigma is implemented through the traditional approach, the Breakthrough approach, or various derivatives, its primary driving factor is the commitment of the company's leader, the CEO. The successful implementation of Six Sigma at Motorola was led by Bob Galvin, and Larry Bossidy. These leaders communicated an enthusiastic vision, a personal desire to achieve results, an expectation of an aggressive rate of improvement, direct involvement in communicating results to stakeholders, and recognition of those who supported the vision. The most fundamental aspect of the vision is superior customer satisfaction that, if achieved, leads to higher profitability. The passionate implementation of the Six Sigma methodology is the means, customer satisfaction is the end, and superior profitability is the financial outcome. Businesses that successfully implement Six Sigma see significant improvement in profitability.

The current Six Sigma approach consists of two implementation levels—the corporate level and the project level. Corporate-level implementation requires leadership to take initiative and middle management to assist in developing a business case for adapting the Six Sigma methodology. They develop the business case by analyzing business performance and identifying factors that adversely affect profitability—in other words, identifying areas where waste of capacity, facilities, funds, and human resources occurs. During this phase, leaders and managers revisit the purpose of the business. They collect supporting data to assess how well the purpose of the corporation is being achieved based on customer feedback, market share, and earnings.

The critical aspects of the corporate-level preparation for the Six Sigma methodology include establishing key business performance measurements, ensuring organizational effectiveness, readying the organization for Six Sigma, and establishing goals for improvement. These goals and opportunities are then aligned with other business initiatives and filtered into projects.

The project-level implementation relies on the DMAIC methodology to capitalize on opportunities for improvement. Extensive training is conducted for champions and sponsors, Black Belt and Green Belt candidates, and employees. The training for champions and sponsors includes an understanding of the need for Six Sigma, Six Sigma's benefits, the rollout plan, the working of Six Sigma projects, the roles and responsibility of all employees (including executives), and an overview of the DMAIC methodology (described below). The Black Belt and Green Belt training programs include various tools and techniques to apply the DMAIC methodology.

DEFINE

The first step—to define—is to clearly describe the problem and its impact on customer satisfaction, stakeholders, employees, and profitability. During this phase, the following are defined:

- Customer critical requirements
- Project goals and objectives
- Team roles and responsibilities
- Project scope and resources
- Process map and Supplier, Input, Process, Output, and Customer (SIPOC)
- Process performance baseline

In understanding the customer requirements, one can learn from Noritaki Kano's quality approach. Kano's approach sorts the customer requirements into three categories: assumed, specified, and expected requirements. An *assumed* requirement is one that's taken as a given. Someone who is buying a car, for example, never checks to be sure that the car will include four wheels. When an implicit requirement is not met, customers are extremely dissatisfied. Assumed requirements, then, are dissatisfiers, where ignorance is the best outcome and loss of the customer is the worst outcome. The *specified* requirements are a customer's explicitly communicated requirements. To meet the requirements is to satisfy the customer. But if an organization has not done anything beyond what customers have specifically asked for, customers will be very open to try competitive products or services. The *expected* requirements include requirements beyond what the customer explicitly communicates. These are the customers' real, unstated expectations of what they would love to receive from suppliers. For a supplier to fulfill these expectations, the supplier must really understand the customers' needs for products or services. The supplier must also anticipate the customers' future needs, thus demonstrating a desire for ongoing relationships with customers, i.e., showing superior customer service.

Once the requirements are understood, they flow down to the operation level, where project goals and objectives are set.

Some of the techniques used in the define phase include the following:

- Project charter
- Stakeholders' commitment analysis

- Affinity diagrams
- Voice of the customer
- Kano's quality analysis
- Force field analysis
- Pareto analysis
- Process mapping
- SIPOC (Suppliers, Inputs, Process, Outputs, Customers)

Some of these tools are used widely in various aspects of a business. However, a couple of these techniques, specifically stakeholder analysis and SIPOC, are used as part of Six Sigma.

The Commitment Matrix (Figure 2-6) is an effective way to assess what support is needed versus the current level of commitment from each of the business functions. An adverse gap between the level of commitment needed and the level of commitment that is currently available identifies the opportunity to increase internal support. Additional support levels can be identified depending upon a company's needs. This analysis of commitment is an excellent way to communicate concern to affected team members. Performing this analysis for each project ensures that necessary support is available to achieve results.

Measure

The purpose of the next phase—to measure—is to describe the opportunity for improvement and quantify the baseline performance. When changes are made for improvement, then the business can verify the effectiveness of the changes. To analyze data, basic statistical techniques such as averages, standard

Commitment	Management	Purchasing	Engineering	Production	Sales
Passionate	Needed		Needed		
Positive		Needed		Needed	
Neutral	Available	Available			
Resistance			Available	Available	
Destructive					

FIGURE 2-6.　Commitment matrix.

deviation, and probability distributions (i.e., the Normal distribution and the Poisson distribution) are critical for understanding the nature of excessive variation in the process.

Variation.   W. Edwards Deming said that "variation is evil." This premise underlies the method for achieving dramatic improvement in any process. Walter Shewhart classified variation as random or assignable. Deming called it common or special variation. The nature of variation depends upon its causes, which could be random or assignable.

Random causes of variation, such as ambient temperature, supplier-to-supplier variation in parts, or operator-to-operation variation, are inherent in the process. *Assignable* causes are those that change for specific reason, such as machine breakdown, an untrained operator, use of the wrong material, an incorrect setup, or some design-related issue. The random causes are difficult to diagnose, and many of them act concurrently, while the assignable causes are known, specific, and introduced in the process. In statistical terms, random causes are the ones that are more likely to happen as a routine (about 95 percent of the time), while assignable causes occur less often (about 5 percent of the time) and are exceptions. In developing statistical thinking, it is not as important to learn many statistical techniques as it is to understand the nature of variation.

Cost of Quality.   Another measure of performance is the *cost of quality*. The traditional cost of quality consists of four categories: internal failures, external failures, appraisal, and prevention. The goal is to increase the preventive cost of quality and to reduce the internal failures, external failures, and appraisal components. Typically, not all the costs of poor quality are measured in a company's accounting system; therefore, it takes both effort to understand and courage to measure accurately the cost of poor quality.

Generally, management's initial reaction is that the cost of poor quality is not significant. For example, a company wants to improve customer satisfaction by reducing the number of defects reaching customers, so it adds inspection and test points. Over time, this inspection and testing becomes a standard process. This process, however, is an activity that doesn't add value to the product. The goal, then, is to reduce the level of inspection or testing as much as possible, as it is a wasteful activity. Figure 2-7 lists some measures of the *cost of poor quality* (COPQ) that must be targeted for reduction.

The objective must be to reduce COPQ and increase investment in the prevention costs. Figure 2-8 shows typical ratios of the cost of poor quality, illustrating that a lot more effort must be committed to improve the prevention cost. Surveys have found that employee training is still the best investment to make in order to add value.

Measurement System Analysis (MSA) is a method used to assess repeatability and reproducibility of the measurement method. The goal is to ensure that the measurement system does not add to excessive variability, thus leading to false conclusions and, therefore, false starts. Normally, it is known as *Gage Repeatability and Reproducibility* (Gage R&R). The Gage R&R can be performed on any measurement method to apportion variances. The measurement method must have resolution an order of magnitude higher than the measurement itself.

Internal Failures	External Failures	Appraisal	Prevention
Failure Reviews	Customer Dissatisfaction	Drawing Checking	Planning
Redesign	Equipment Downtime	Final Inspection	Capability Studies
Reinspection	Excess Inventory	In-Process Inspection	Design Reviews
Repair Costs	Excess Travel Expense	Laboratory Testing	Field Testing
Retesting	Excess Material Handling	Personnel Testing	Vendor Surveys and Evaluation
Rework	Penalties	Receiving Inspection	Procedure Writing
Scrap Allowances	Pricing Errors	Product Audits	Training
Engineering Changes		Shipping Inspection	Market Analysis

FIGURE 2-7.   Cost of poor quality (COPQ) measures.

COPQ Category	Estimated Contribution, %
Internal Failures	25 – 40
External Failures	25 – 40
Appraisal	10 – 50
Prevention	0.5 – 5

FIGURE 2-8. Cost of poor quality (COPQ) contributions.

Six Sigma Measurements. A *defect* is defined as any attribute of a product that does not provide total customer satisfaction. The Six Sigma methodology measures defects in two key ways: *Defects per Unit* (DPU) and *Defects per Million Opportunities* (DPMO). A *unit* is defined as the output of a process. For example, a unit for the accounts payable department may be an invoice; for the assembly area it might be a subassembly; and for the packaging department, it may be a package that will be shipped to the customer. The DPU can be calculated as follows:

$$DPU = \frac{\text{Total number of defects}}{\text{Total number of units inspected or verified}}$$

The DPU measurement uses total defects instead of total defective units. For example, when a cell phone is inspected and five defects are found, all defects must be counted, recorded, and included in the DPU calculation. Once the DPU is calculated, the first-pass yield can be calculated according to the following formula:

$$\text{First-pass yield} = e^{-DPU}$$

Normally, the yield is calculated according to the number of good units produced over the total units started. If utilized correctly, the yield number will appear more accurate. However, because the focus for improvement is on defects or errors, the number of defects must be measured; they point to the opportunities for improvement.

When using measurement or variable data, one needs to look at the distributions of data and utilize an appropriate distribution to determine probabilities of producing good product within specifications. The following steps can be used to predict yields based on the variable data:

STEP 1. Gather variable data.

STEP 2. Calculate the average and the standard deviation.

STEP 3. Calculate the probability of producing the product within specification, using the Normal distribution table (in any business statistics book or software).

STEP 4. Add probabilities of producing the product within specification on both sides of the target.

STEP 5. Subtract from 100 to determine the defect rate. The defect rate can be converted to parts per million.

For example, if the process, as shown in Figure 2-9, demonstrates standard deviation such that the 3 Sigma distance is equal to the specification limits, then the 99.73 percent process output will be acceptable. If, however, the standard

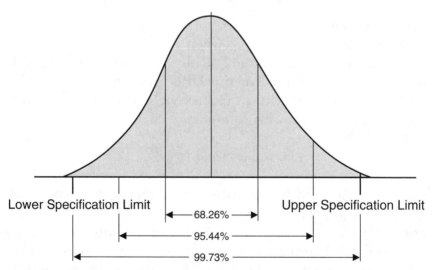

FIGURE 2-9.   Probability of producing products within limits at specified standard deviations.

deviation is reduced through process improvement, or the limits are opened up after negotiations with customers, such that the tolerances are higher than the 3 Sigma around the mean, then the predictable yield will be 99.9 percent or higher. It is at this point that the benefits of reduced inspection and testing become visible.

ANALYZE

During the analyze phase, the focus is on searching for the root cause. Based on the data analysis, opportunities are prioritized according to their contribution to customer satisfaction and impact on profitability.

Pareto Analysis. The *Pareto chart* (see Figure 2-10) is a graphic representation of the opportunities for improvement. It is used in identifying the critical opportunities that will have the greatest impact on customer satisfaction and profitability. The chart was developed by J. M. Juran and named after the Italian economist Vilferado Pareto, who observed that most of the world's wealth is owned by a few individuals. Juran likewise

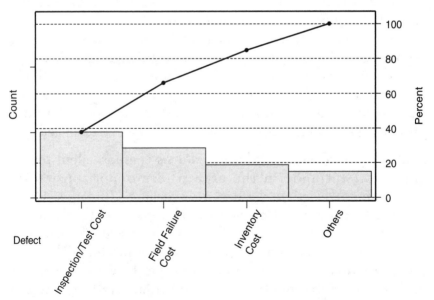

FIGURE 2-10.   Pareto chart of opportunities for improvement.

found that most events in nature are not equal. For example, most revenues of a company come from a few large accounts, most of the deaths occur due to a few diseases, and most problems in an organization stem from just a few causes.

The Pareto chart is designed to help businesses identify opportunities for improvement that cost more than the others and thus should be attacked first. The Pareto chart shows opportunity categories based on their impact or frequency. People tend to work first on opportunities that are easier to attack rather than those most important to attack. The purpose of using the Pareto chart is to promote work on important opportunities rather than on the easy ones.

Cause-and-Effect Analysis.   Once the most significant opportunities have been identified, a root cause analysis is performed. The cause-and-effect diagram is one tool used to diagnose the causes of a selected problem. Most failures are caused by problems with the machine, material, method, or mind (skills). In addition, the environment and measurement devices may cause failures. The cause-and-effect diagram is a great way to list potential causes. Once causes are listed, a cross-functional team can prioritize various causes and select a few on which to work. The cause-and-effect diagram is also called the *Fishbone* or *Ishikawa diagram.*

As shown in Figure 2-11, the main branches can be relabeled according to the categories of causes to be investigated. In the case of financial losses, categories such as machines, methods, and materials might not be the appropriate categories to represent potential causes. In such cases, other causes can be fitted into the standard branches, or the branches can be relabeled.

Multivary Analysis.   *Multivary analysis* is an excellent tool to apportion variance in the area where opportunities for improvement exist. It dissects the variance into positional, cyclical, and temporal categories. The positional variation is caused by the variables that affect the process performance at certain locations within the process or the product. The temporal variation is attributed to the changes between cycles of a process and represents trends over time, i.e., shift to shift, day to-day, or week-to-week.

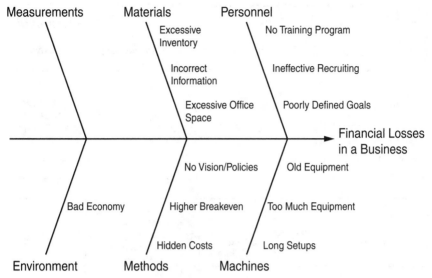

FIGURE 2-11.   Cause-and-effect diagram.

Apportioning the variation in a process puts focus on the variables related to a particular type of variation. For example, the positional variation is normally attributed to product or process design, as the defects recur at certain locations. The cyclical variation is attributed to the variables related to the process setup that cause variation in performance from one process cycle to the next. The temporal variation can be related to the maintenance activities, whether daily, weekly, or monthly, as well as degradation in consumable items in the process, such as laser lamps, machine tools, chemical concentrations, or the limited shelf life of a chemical.

FMEA. *Failure Mode and Effects Analysis* (FMEA), as shown in Figure 2-12, is an excellent tool that has been used in mainly the automotive and aerospace industries, or where personnel safety is a concern. As implied by the name, FMEA is used to anticipate potential failure modes during the product or process design or redesign, to determine the effects of failure modes on performance, and to identify action items that will prevent anticipated failure modes. Each failure mode is ranked for severity of the effect on performance, frequency of

					Description of
					Protection: The spreadsheets

System _____

Subsystem _____

Component _____

Design Lead _____

Core Team _____

Potential Failure Mode and (Design

Key Date_____

Item / Function	Potential Failure Mode(s)	Potential Effect(s) of Failure	S e v	Potential Cause(s)/ Mechanism(s) of Failure	P r o b	Current Design Controls
Coolant containment. Hose connection. Coolant fill.	Crack/break. Burst. Sidewall flex. Bad seal. Poor hose retention	Leak	8	Over pressure	8	Burst, validation pressure cycle.
	Write down each failure mode and potential consequence(s) of that failure.					
	Severity - On a scale of 1–10, rate the Severity of each failure (10 = most severe). See Severity sheet.					
				Likelihood - Write down the potential cause(s), and on a scale of 1–10, rate the Likelihood of each failure (10 = most likely). See Likelihood sheet.		

FIGURE 2-12.   Failure mode and effects analysis template.

FMEA Worksheet

are not protected or locked.

Effects Analysis
FMEA)

FMEA Number _____

Prepared By _____

FMEA Date _____

Revision Date _____

Page _____ of _____

D e t	R P N	Recommended Action(s)	Responsibility and Target Completion Date	Actions Taken	Action Results			
					New Sev	New Occ	New Det	New RPN
1	64	Test included in prototype and production validation testing.	J.P. Aguire 11/1/95 E. Eglin 8/1/96					

Response Plans and Tracking

Risk Priority Number - The combined weighting of Severity, Likelihood, and Detectability. RPN = Sev X Occ X Det

Detectability - Examine the current design, then, on a scale of 1–10, rate the Detectability of each failure (10 = least detectable). See Detectability sheet.

FIGURE 2-12. (*Continued*)   Failure mode and effects analysis template.

occurrence of its cause, and detection of the failure mode based on the effectiveness of the control methods. A *risk priority number* (RPN) is calculated by multiplying the ranking for severity, occurrence, and detection. The RPN is used to prioritize the failure modes and corrective actions related to the failure modes.

Improve

The improve phase consists of developing solutions and selecting the optimum solutions for best results and most robust performance. Two key aspects of the improve phase include the use of *Design of Experiments* (DOE) and change management.

The traditional approach to finding a solution to a problem focuses on one variable at a time, holding the other factors constant. Shortcomings of this approach include the following:

- It is usually not possible to hold all other variables constant.
- Too many experiments are required to study the impact of all the input variables.
- The interaction between variables cannot be determined.
- The optimum combination between variables may never be revealed.
- Resources might be wasted in studying the wrong variables.

Statistically designed experiments involve varying two or more variables simultaneously and obtaining multiple measurements under the same experimental conditions. The objective of DOE is to assess the effects of the critical variables and the interaction among them, and then to determine the significance of those effects compared to the experimental error. If the effects of the process changes happen to be significantly better, a new process can be implemented.

The advantages of this approach are the following:

- Many variables can be measured simultaneously, making the DOE approach more economical.
- Interactions between variables can be detected and measured.

- Experimental error is quantified and used to determine the confidence level in the experiment results.
- The results will indicate if any important variables were missed in the experiments.

To improve a process, one must gain knowledge of the process, its environment, its components, and its responses. The following steps outline the various steps for process improvement:

- Describe the process and break it into subprocesses.
- Determine measurement instrument variability.
- Collect and analyze data.
- Identify major components of variation, using statistical tools.
- List input variables associated with the major component of the variation. Screen out trivial variables by using brainstorming, screening experiments, or similar techniques.
- Conduct statistically designed experiments to identify critical input variables.
- Optimize the process with attention to critical variables, and determine realistic tolerances of input variables.

Figure 2-13 shows a flowchart that describes the use of various techniques deployed in the design of experiments. The Full Factorial technique, commonly used for most process improvement activities, requires selection of the key variables, the desired settings for the new process, and the amount of improvement desired. Once the experiment is designed, the experimental cells are randomized, and the data are collected and analyzed using a statistical software. Most experiments provide additional knowledge about the process and lead to convergence in solving the problem. Sometimes multiple experiments are required to arrive at a solution.

CONTROL

Once the improvement is realized, the goal is to control the improved processes and sustain the Six Sigma initiative. Tools

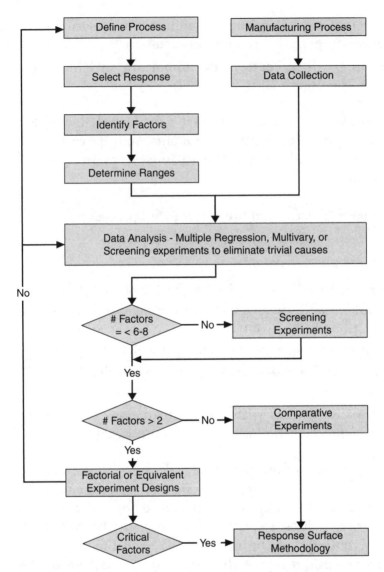

FIGURE 2-13. Strategy for conducting design of experiments.

such as control charts, precontrol charts, and run charts are used to maintain the processes. The challenge, however, lies in keeping the Six Sigma initiative alive on a continual basis.

One way to keep the Six Sigma initiative alive is to have the organization divide itself into functional territories, with managers in each territory working only on improving their own functions. This solution, however, is sometimes detrimental to the organization as a whole. Such an organizational hierarchy can give managers a monopoly for controlling any changes to their departments or functions. This authority can lead to resistance to any "outside" interference and change, such as quality improvement projects. On the other hand, some organizations have too many rules (or written procedures), and any change in those rules must go through a bureaucratic process, which may take weeks or months. To lower such barriers, incentives to drive dramatic improvement or consequences for resistance must be clearly defined and communicated.

Leadership must transform the organization's culture into one that embraces change rather than fights it. A mentor can help project managers find the right resources or address critical roadblocks. The mentor may direct the project manager to an informal leader whose support is critical for the project's success as well as find legitimate ways around constraints and roadblocks. For example, a minor change in the project plan might help bypass a daunting approval requirement.

CHALLENGES IN IMPLEMENTING SIX SIGMA

Sometimes, a company's leaders adopt the Six Sigma initiative without understanding the potential opportunities that can increase profitability. They basically treat it like another cannonball or another flavor of the month. While these companies have committed significant resources in training people for Black Belt and Green Belt certification, either the projects for the trained personnel to work on have not been identified, or the personnel have not been allowed to work on the projects because of other priorities in the company.

At a large automotive original equipment manufacturer, the Six Sigma methodology is treated as a five-step problem-solving methodology, and management does not get involved. At another company, projects are already in progress before the Six Sigma training. After the Six Sigma training, the projects are counted toward the Six Sigma success. At another company, a consulting company asked the project leader to fudge the numbers to make savings look better than they were. At an American Society for Quality section meeting, I was told that the company had just completed the Six Sigma training, but that it may need some help in project selection. These examples are all start-up problems with the Six Sigma methodology's implementation.

At some companies, the challenge has been to sustain the rate of improvement and successes over the long haul. Problems have occurred because of misalignment of corporate strategies and ineffective performance management. The most significant cause of poor performance management is unavailability of the proper feedback at all levels, i.e., from the process all the way to the leadership. The current available measurements may be limited to sales reports, financial measures, and the performance of production processes. Several areas in the corporate world have been ignored. As discovered through the Six Sigma methodology at Motorola and GE, significant savings can be realized in service areas as well as in production. Therefore, performance measurements at key processes throughout the organization are crucial.

Another weakness in some organizations, whether they are implementing Six Sigma or other similar methods, is the hesitancy of leaders to set goals for an aggressive rate of improvement. Aggressive goal setting is a critical characteristic of the Six Sigma methodology, as it demands innovation in each area, intellectual involvement of all employees, and dramatic improvement. Minor adjustments in production or management processes lead only to incremental improvement that is often hidden by the error levels in the corporate management system.

To share the benefits of the improvement resulting from application of the Six Sigma methodology, the rate of improvement must be greater than the rate of inflation and measurement errors. With the changing corporate competitive environment, implementation of an ISO 9000 system at more than 500,000 companies, acceptance of Six Sigma methodology for superior performance, and the application of Malcolm Baldrige National Quality Award guidelines worldwide, a comprehensive, practical and progressive performance system is needed to realize dramatic improvement on the order of 50 percent. The Six Sigma Business Scorecard combines various performance measures and goal-setting methods of Six Sigma to meet corporate needs for dramatic improvement and higher profitability.

TAKE AWAY

1. Motorola discovered that customers expected better products than they were getting and that problems in the field were extensions of internally observed problems.

2. Typical accounting practices do not consider all aspects of the business, as the focus is on the visible waste. The hidden costs need to be understood, identified, and monitored in order to maximize an organization's profitability.

3. The implementation of the Six Sigma methodology means total customer satisfaction, and superior profitability.

4. DMAIC stands for Define, Measure, Analyze, Improve, and Control.

5. Define phase clearly describes the problem and its impact on the customer satisfaction, stakeholders, employees and profitability. Key tools and techniques include Project Charter, Voice of Customer, Kano's Quality Analysis, Force Field Analysis, Pareto Analysis, Process Mapping, and SIPOC.

6. Measure phase describes the opportunity for improvement and quantifies the baseline performance. When changes are made for improvement, the business can verify the effectiveness of the changes. Key tools and techniques include Cost of Quality, DPU, DPMO, and descriptive statistics in terms of mean and variance.

7. Analyze phase focuses on searching for the root cause. Key tools and techniques include Cause and Effect Analysis, Multivary Analysis, and FMEA.

8. Improve phase consists of developing solutions and selecting the optimum solutions for the best results and the most robust performance. Key tools and techniques include DOE and Change Management.

9. Control phase sustains the improved results achieved during the improve phase. Key tools and techniques include Control Charts, Precontrol Charts, and Run Charts.

10. Key challenges in implementing Six Sigma include adopting the Six Sigma initiative without understanding the potential opportunities that could increase the profitability, not identifying the projects before investing the resources, misaligned corporate strategies, ineffective performance management, and reluctance to set aggressive improvement goals.

NEED FOR THE SIX SIGMA BUSINESS SCORECARD

A recent business assessment survey [developed by the National Institute of Standards and Technology (NIST)] and conducted by Quality Technology Company led to some interesting findings about business performance. Most respondents reported that they have no idea how company leadership planned to improve the profitability. Only a few people knew how successfully their department was in achieving business objectives.

The respondents in the survey represented owners, managers, engineers, and operators of several companies (including those in the Fortune 100). The responses of employees to questions in various business areas are summarized in Table 3-1.

TABLE 3-1. Business Assessment Survey Findings

BUSINESS AREAS	EASY TO ANSWER	COULD ANSWER	DIFFICULT TO ANSWER
Organization Environment	53	38	9
Organizational Relationships	32	46	22
Competitive Environment	27	41	32
Strategic Challenges	26	30	44
Performance Improvement System	19	37	44
Overall	**32**	**38**	**30**

SOURCE: www.NIST.org.

As is apparent from this table, only about one-third of the employees can easily answer questions about business performance, implying that information about the business performance is not readily available to employees.

Interestingly, the consistency of responses among the sections was quite surprising. The two weakest areas are the strategic challenges and the performance improvement system. In the case of the process improvement system, only about 20 percent could easily answer any questions. (See Table 3-2 for details on the process improvement area.)

TABLE 3-2. Response to Process Improvement–Related Questions

PROCESS IMPROVEMENT–RELATED QUESTIONS	EASY TO ANSWER	COULD ANSWER	DIFFICULT TO ANSWER
How do you maintain an organizational focus on performance improvement?	26	23	61
What approaches do you use to systematically evaluate and improve key processes and to foster organizational learning and knowledge sharing?	9	41	50
How do you measure product and process quality? What processes do you have in place to evaluate root cause and improve performance?	21	49	30
At what rate is your operations performance in quality, cost, cycle time, and profitability improving annually?	6	30	64
How do you maintain an organizational focus on customer satisfaction? What processes do you have in place to improve customer satisfaction?	24	43	33
How do you evaluate employee job satisfaction? What processes do you have in place to initiate steps to improve employee morale?	15	26	59
Overall	19	37	44

MISSING RATE OF IMPROVEMENT

The most significant finding is that people are unaware of the rate of improvement in their companies. They are also unaware of their process for improvement, implying that the focus is on delivery, not on getting better. In addition to reporting being unaware of process improvement, most respondents indicate that mechanisms to assess employee satisfaction are very sparse. Only 15 percent of people could easily answer questions about their company's process for improving employee satisfaction. Considering that employees' input is critical to the leadership, for generating new ideas as well as for getting feedback about the performance of the company leadership, this finding is troubling. As for the competitive position of the company, only a few employees know about their competitors and their competitive advantages. This lack of awareness can limit a company's improvement.

Employees are very familiar with their customers, as well as what their customers want. At least they think so! In the current business environment, the supply chain relationship plays a critical role in the success of a company; employees, however, have limited interaction with their suppliers. Supplier relationships have become very important as businesses become increasingly dependent on their suppliers.

These employees know what they are supposed to do, but they have a difficult time figuring out how well they are doing. Whether they are in engineering, sales, purchasing, production, quality, or another department, they do not have measures of performance for the various aspects of their jobs. They believe that their company has measurements for the short term but does not care about long-term performance. They have measurements about the function of the company, but they do not have the measurements about the company's strategic intent. Employees are unaware of the competitive position of their products and services, mission and values, and results achieved. In summary, employees feel good about their products and services; however, they are not informed of the performance levels expected or what rate of improvement they are supposed to achieve.

INSUFFICIENT PROCESS PERFORMANCE MEASUREMENTS

Most businesses have measurements for sales and profitability. They do not, however, have measurements for operational effectiveness. Every company has an accounting function, whether internal or outsourced, to summarize the accounts payable and receivable, balance sheet, and profit and loss. Many organizations, although they hope to be profitable, are penny-wise and dollar-foolish.

Some companies have established measurements that support the premise that they are profitable. This premise is quite possible in industries with a niche market, a lack of competition, good margins, and some waste that can be tolerated. However, that privilege is only for the lucky few. Most companies have stiff competition, complex products, and a complex system that does not identify opportunities to reduce waste.

A typical business today consists of buildings, equipment, employees, material, customers, and management. The organizational structure may include one or multiple facilities at one or several locations. Besides that, a business may have several parts, products, or types of services that it offers to customers.

Typically, the business is managed by making sure products are delivered to the customer in the best possible manner. The focus is on shipping the product or delivering the service as best as possible to maximize the revenue. At the end of the month, the leadership knows how much money has been made in profits or how much more credit is needed to manage the cash flow.

Inside a company, life is a little more interesting. First, the sales staff are busy selling more to earn a commission than to make margins for the company. For example, a salesperson makes a deal for $500 million and earns a commission. The company finances the sales and pays out the commission. The revenue is recorded on the books, showing sales growth. A few months later, however, the customer files for bankruptcy. The company ends up in debt for $500 million, while the salesperson already earned the bonus and recognition and moved on.

Meanwhile, design engineering is busy designing the part or the product. The product is designed to best-of-design capabilities, which are practically all imaginary and incorporating all the creative juices and genius of the design team in each product. Some prototypes are built and tested for functionality. The design works by hook or crook, and the product is released to production. Production may pilot-test a few runs to verify the design. When the product works, it is signed off for full production and shipment to the customer.

Here is where the fun begins. Purchasing is buying parts from the suppliers. Suppliers celebrate their success in making sales to the company. Production, however, is having trouble producing a good product. Sometimes the product fails due to bad parts; sometimes it is incorrectly assembled by machine operators who are tired of fighting fires. Sometimes failure is due to marginal design specifications; at other times inspection and testing did not catch enough failures. In addition, the product fails at times in the field because the company CEO told employees to ship the product at any cost in order to keep his word to his customers (even if it meant skipping some verification steps).

Now the customer returns the product. Someone receives it in the company, reviews it, and assigns credit to the customer. Someone, if any time can be found, analyzes the root cause of the problems and finds an operator who needs some counseling. Quality engineering issues a corrective action, completes the paperwork, and closes the corrective action. The customer gets a copy of the completed corrective action, accepts the response, completes paperwork, and life goes on.

In addition to this life in sales-to-service operations, other business activities are ongoing. Managers make sure the product gets out the door. The support staff expedite materials and whatever else is needed to keep the operations running. They ensure bills are paid and invoices issued, that employees are paid, and that struggling employees get help when necessary.

The most interesting personalities may be those of the owners, CEOs, or presidents. They are earning a good salary, and their companies have been having some profitable years due to favorable market conditions. They are stressed, though, because they do not understand what is going on in the real business. They ask the staff how things are going, and the staff respond, "Just fine!" They do not probe further, because they believe that is the best answer they can get. Digging any deeper might stir debate, and they want to avoid conflict with the staff, even if it means avoiding the truth.

These situations may typically occur in a privately owned and improperly staffed small business—one run by an old-style management. However, similar situations occur in divisions of large corporations as well. They happen because those in charge lack understanding of the business, fail to identify a common purpose, and cede management of the processes. As a result, each employee is trying to manage his or her time at work independently. Many inefficiencies result from conflicting priorities and questionable interests on the part of employees.

Some companies have a few performance measures in place, even fancy-looking reports and charts. However, their basic goal seems to be to produce parts as fast as possible. Keep everybody physically busy instead of allowing time to be mentally busy.

Many businesses have very few measurements; some have a lot of measurements but not necessarily the right ones; others may have extensive measurements imposed from headquarters. However, these measurements do not correlate with the business reality because of ineffective communication between leadership, management, and staff. The result is insufficient review and analysis and an improper focus on speed instead of doing the job well.

Henry Ford once said that a waste of time is much more significant than a waste of material, because the material has some salvage value, while time has none. Wasted time is a particular problem for many service industries because so much of the work they do is based on intangibles, giving them more "wiggle" room. Yet service industries face similar issues in terms of measurements and performance. The issues of learning to perform better and faster at lower cost in the service industry are identical to the ones faced by the manufacturing industry.

In larger corporations, the complexity and diversity of operations mean that the overhead to arrange a meeting costs more than the meeting itself. In one company, I asked my next-door employee to set up a meeting time, as I was new there. He gave me a date for an hour to meet but a month out. There was no sense of urgency to get the job done at this corporation, and the time spent scheduling and waiting for meetings that should take place quickly wasn't recognized as a cost.

DEMING'S 14 POINTS*

All this discussion points to the conclusion that something is missing in our businesses. Deming, who was credited with jump-starting Japan's economy to its heights, identified 14 organizational transformation points several years ago:

1. Create constancy of purpose toward improvement.
2. Adopt the new philosophy.
3. Cease dependence on inspection to achieve quality.
4. End the practice of awarding business on the basis of the price tag.
5. Improve constantly and decrease costs.
6. Institute training on the job.
7. Institute leadership.
8. Drive out fear.
9. Break down barriers between departments.
10. Eliminate slogans, exhortations, and targets for the workforce.
11. Eliminate work standards (quotas) on the factory floor and management by objectives.
12. Remove barriers that rob the employees of their right to pride of workmanship.
13. Institute a vigorous program of education and self-improvement.
14. The transformation is everybody's job.

*W. Edwards Deming, *Out of Crisis*, Knoxville, Tenn.: SPC Press, 1982.

Deming's 14 points were well recognized in the late 1980s and early 1990s. Each point communicates a strategy that, if implemented, will transform an organization. These 14 points promote improvement through leadership and the system instead of through management by objectives. The first point about creating constancy of purpose is the most critical one. The leadership must be driven to maintain the continuity of the business in a competitive environment by creating jobs, growth, and profitability. The last point emphasizes empowering employees through their involvement in decision making and active participation from planning to performance.

If we look at all 14 points in their entirety, it appears that Deming's approach is to create an organization that is natural in its functioning. In other words, the organization recognizes the intelligence of all employees, the uncertainties associated with people, superior leadership, performance of processes instead of people, and effective measurements. Deming believed that leadership's role must be to care for employees, develop their skills, and achieve business objectives through employees' total involvement. Deming's main concern was the mobility of leadership and disruption caused by this mobility. Leadership's lack of commitment to transforming an organization into a profitable one over the long term is a leading cause of poor performance.

MEASUREMENT CHALLENGES WITH QUALITY SYSTEMS

LIMITS OF ISO 9000

The ISO 9000 standards process model is depicted in Figure 3-1. According to the process model, any activity, or set of activities, that uses resources to transform inputs to outputs can be considered a process. For a company to effectively implement a quality management system, it needs to identify and manage various processes and their interactions. The output of one process is an input to another process. To implement a

FIGURE 3-1. Process-based quality management system. (*ISO 9000:2000.*)

companywide quality management system, the company must implement process management at each process (function or subsystem) level.

To manage a process, the process owner needs to control inputs, in process, and output as shown in Figure 3-1. The process owner must not only understand the requirements but also be able to receive, produce, or supply according to the requirements. Verification methods are needed to ensure compliance to the requirements at various stages of the process. This might involve controlling suppliers or monitoring the product or process through data analysis, inspections, testing, or measurements. If the verification shows that requirements are not met, the process owner needs to initiate corrective action.

Once we learn how to manage a process effectively, we must identify critical process steps throughout operations, verify them, ensure that they comply with requirements, and then produce acceptable results. According to ISO 9001:2000, a process is run effectively when the process owner ensures the following: processes are clearly defined and documented, steps are clearly documented, employees are trained, good data are collected, data are analyzed, and corrective actions are taken.

If the roles and responsibilities of every employee are defined and documented and accountability is established, company management can smoothly run the business with the process management mentality. Of course, if it's not monitored, the quality management system can be blamed for any problems. With process management, the difference between employees and management disappears. Everyone, including management, must perform the assigned tasks according to the intent and processes of quality management. With this synergy and harmony, the company develops an effective quality management system that can facilitate growth as well as downsizing equally well. A well-run quality management system should make the company perform like a well-oiled machine.

Today, about 500,000 businesses have invested heavily in the ISO 9000 system. The ISO 9000 system allows a company to create an infrastructure that guides process management throughout the organization. The most common benefits of implementing ISO 9000 quality systems include the following:

- Consistency, standardization, and repeatability of processes
- Increased business
- Customer confidence
- Better management and less confusion in the plant

There are challenges with the system, however, that have led to suboptimal performance. Some of the industrywide concerns include

- Excessive documentation or paperwork
- No change in the way of doing business
- Management ignorance
- Invisible improvement in performance

The challenges imply the lack of a performance measurement system that maintains the accountability between leadership and the operator. In other words, some roles are not played well—typically, the nonproduction roles.

SHORTCOMINGS OF SIX SIGMA MEASUREMENTS

At the same time as the ISO 9000 standards were released, the Six Sigma initiative was launched at Motorola. Some companies that have implemented Six Sigma have had difficulty sustaining the optimum level of performance. Each successful Six Sigma implementation was passionately led by the company's chief executive. Some of the key aspects of successful Six Sigma implementation include the following:

• Commitment
• Accountability
• Aggressive goal setting
• Communication
• Common language
• Process thinking
• Innovation
• Metrics
• Rate of improvement
• Rewarding experience

To ensure that these key aspects are implemented successfully, measurements are needed to monitor progress. Typical measurements include the following:

• Defects per unit (DPU)
• Defects per million opportunities (DPMO)
• Process yields
• Customer satisfaction
• Customer returns
• Employee suggestions

It appears that Six Sigma measurements focus on performance at the process level; however, the measurements are not aggregated and correlated to corporate wellness. Corporations have found it difficult to establish a corporate sigma level that correlates with the overall corporate performance.

LIMITS OF THE BALANCED SCORECARD

Current measurement systems tend to focus on operations, and measurements for the strategic aspects of the business are limited. This leaves leadership unable to relate to the overall performance of the business. Robert Kaplan and David Norton (1996) addressed this discrepancy when they developed the *Balanced Scorecard* in the early 1990s. The idea was to supplement the usual financial measures that were insufficient to manage modern organizations. Their view was that a balanced group of measurements that supports the corporate strategy would enable an organization to achieve the business objectives. A Balanced Scorecard includes measures in four areas: Financial, Customer, Internal Business Processes, and Learning and Growth, as shown in Figure 3-2. The Balanced Scorecard reveals a much broader view of what is happening in an organization than traditional financial measures alone do. This broader view, though, is only part of the value added

FIGURE 3-2.   Balanced Scorecard system. (*Kaplan and Norton, 1996.*)

by the Balanced Scorecard approach. The real contribution of a Balanced Scorecard program is to link the objectives in each of the four perspectives. Each organization selects specific measures and draws specific links between them.

The Balanced Scorecard is best deployed at the strategic level and flowed down through the organization. Work groups can devise their own Balanced Scorecards that show their contribution to the strategy of the organization. Action plans and resource allocation can be determined according to the work groups' contributions to the corporate Balanced Scorecard objectives.

While implementing a Balanced Scorecard, managers articulate their strategy for the organization. Departments go through the training and attend sessions to develop the vision, strategy, and measurements that will lead to a Balanced Scorecard. They develop objectives and targets as well as action plans. Weaknesses in the organization can be identified through the reporting process and corrected through the learning process.

In theory, if a Balanced Scorecard is created at each department level, it could become a major measurement challenge. People who have experienced the process, however, say that by the time the Balanced Scorecard gets to work groups, the strategy has become unrelated to employees and too much effort is required to maintain the system. As with any new methods and their associated learning curves, challenges are expected.

The Balanced Scorecard has been successfully implemented at hundreds of companies; however, the remaining millions of businesses still need a practical measurement system that will enable them to improve profitability. As Kaplan and Norton state in *The Strategy Focused Organization* (2001), the execution of the measurement system is more important than the measurement system itself. Accordingly, fewer than 10 percent of the strategies outlined on the Business Scorecard were successfully implemented. This implies that the measurement strategy must be simplified for a successful execution.

MEASUREMENTS ACCORDING TO
THE GOAL

Another classic approach, developed by Eliyahu M. Goldratt and Jeff Cox in the book *The Goal* (1992), appears more practical in the sense that it simplifies the main strategy for a business: *to make money*. The key measurements suggested in *The Goal* are Throughput, Inventory, and Operating Expenses. This approach stresses two main underlying beliefs: (1) Each business has many constraints, and (2) a linear approach to improving profitability sometimes leads either to localized improvement but no impact on profitability or to an adverse impact on profitability. Typically, businesses measure things such as shipments, income, speed, behaviors, work environment, physical space, overtime, efficiency, competition, market, technology, quality, scheduling, delivery, constancy of purpose, expenses (short-term and long-term), loss of good employees, sales, bottlenecks, productivity, effective meetings, R&D effectiveness, return on investment, and cash flow. These measurements tell us how various aspects of the businesses are performing; however, they do *not* tell us whether they are adding to the bottom line.

DEVELOPING A NEW MEASUREMENT SYSTEM

What an organization truly needs is measurements that start with its objectives and relate to processes that can be aggregated to overall performance. When this happens, measurements directly relate to profitability. In other words, the measurement system starts with a goal, it flows down to the process level, and it can be optimized. The measurement system must be able to provide a prompt and futuristic snapshot of the business's health to the executives.

A measurement system itself should not be a major expense. The ideal system can be quickly summarized, reported, communicated, and acted upon to prevent opportunities for waste of resources, including intellect. A good measurement system promotes intellect rather than effort.

For example, a company of about 150 employees wanted to implement an improvement process to reduce waste and improve profitability. The initial attempt to develop a well-defined strategy with goals, objectives, targets, and a sound plan for execution failed. The plan was impossible to implement due to the required effort, existing culture, and skills of the employees. The company's leadership then changed tactics. A new system was devised in which (1) each area supervisor was given a continual improvement goal, (2) a weekly review was performed, and (3) appropriate feedback was given on a green sheet or a red sheet. To our surprise, the supervisors were able to improve processes in 44 out of 52 weeks and reduce waste by about 80 percent. The company experienced its best-ever operational and financial performance.

Soon after, a new leadership team lands in the company. The leaders decide to bring their cronies and implement their own system. Soon enough, the company loses all the benefits of its previous system, loses money, and eventually closes.

Time after time, it is not the company's processes or measurements that are the problem; rather it is that *leadership* is not defined as a process that should be measured for maintaining constancy of purpose, direction, and profitability. In other words, a good measurement system must include measurements for leadership as well.

Profitability is a real challenge in today's business environment due to increased competition, reduced development time, shrinking margins, and pressures on price. To achieve increased profitability, objectives must be clearly defined and communicated, and the system must be optimized. The measurement system must provide feedback to fine-tune the system so that the business can achieve its profitability objectives— just as a guided missile that, with some statistical variation, hits its target. With the help of the right measurements and proper execution, the accuracy of the business system to hit the target must be refined for long-term sustainability.

Typically, leadership has focused on sales, finances, and the operational measurements of quality and cycle time. The more recent Balanced Scorecard measurements focus on financials, learning, internal processes, and customers—all linked to the strategic level. The Balanced Scorecard measurements, however, have not worked effectively at the operations level; thus, they are not suitable to a business's organizational structure.

A new business scorecard system based on the knowledge gained thus far through various efforts is called for—one that not only directs the organization to achieve profitability, but also maintains profitability by ensuring a high rate of improvement in internal operations. The level of performance is not what counts; instead the *rate* of improvement is what matters to achieve profitability. The Six Sigma methodology, if executed as intended, achieves improvement objectives, and the Six Sigma Business Scorecard helps to achieve the intended profitability objectives.

The Six Sigma Business Scorecard combines information from the strategic, operational, and execution aspects of the business. Only through aggressive goal setting, effective data collection, analysis, reporting, communication, and improvement efforts can a business achieve the desired objectives. A Six Sigma Business Scorecard reflects the effectiveness of all processes at the action level instead of the strategic level. The Six Sigma Business Scorecard relates to the organization instead of viewing the organization from the measurements perspective.

TAKE AWAY

1. Most employees have no idea how company leadership plans to improve the profitability.

2. Most people are unaware of the rate of improvement in their companies. They are also unaware of their process for improvement, implying that the focus is on delivery, not on improvement.

3. Employees' lack of awareness of their competitors and competitive advantages can limit a company's rate of improvement.

4. Each employee is trying to manage his or her time at work independently with the focus to stay physically busy instead of being mentally active.

5. Fewer than 10 percent of the strategies are successfully implemented. The execution of the strategy is equally important that can be facilitated by effective performance measurements.

6. The measurement system must provide a prompt and futuristic snapshot of the business's health to the executives. A good measurement system promotes investment rather than cost to fight fires.

7. With the help of the right measurements and proper execution, the accuracy of the business system to hit the target must be refined for sustained profitable growth.

8. The Business Scorecard coalesces the strategic, execution, and innovation aspects of the business.

THE SIX SIGMA BUSINESS SCORECARD

Businesses have one target—to be profitable. But the main purpose of a business is not to make money; instead, it is to provide products or services that its customers want. If a business does that well, profits will follow. Unfortunately, that's easier said than done.

Successful businesses must grow and be profitable. If they do not, they are using society's valuable resources and putting them to a purpose with no value. This is wasteful and wrong. Each business's objective should be to provide a product or service that is within its core competency, take care of its customers, plan to develop its people, and be profitable for shareholders.

BUSINESS AND LEADERS

At the initial stage of business development, limited product and services are offered, the focus is on a few customers, and performance is easily visible. However, as the business becomes successful, its leaders generally plan to grow or diversify. Leadership identifies new market segments, develops new products or services, recruits additional people, acquires more equipment, and secures consultants to help plan strategy. The business becomes complex, as processes become numerous, material flows through a convoluted network, cash flow becomes questionable, and debt increases. Performance becomes less and less visible. Leadership continues to plan for success, but profits become more difficult to acquire. Although there are many exceptions, this is a typical path.

Why is it that the same people who succeed early on in a business later find it difficult to maintain their successful profile? Have these people simply reached their level of incompetence, as business management gurus suggest? I am not sure if that is the case. Successful leaders are similar to successful businesses. They have developed a formula for success. In other words, they have optimized a good business system that takes specific inputs and transforms them to the necessary output. Customers like their products or services and pay well for them, and the business is successful. But the leadership is unable to sustain the same level of success indefinitely. What really happens is that as certain aspects of the business change, new variables are introduced and the business system loses its optimum level of performance. Challenges show up and profitability evaporates.

To regain profitability, businesses need to relearn the business system, optimize it, and make it profitable. Sometimes new leaders with open minds turn the business performance around; other times reenergized leaders work to change the system to improve performance.

Basking in the glow of recent successes, some business leaders look for a new opportunity for advancement. As they change companies, two business systems are disturbed—the one they left and the one they have entered. Businesses that are successful over the long haul have dynamic leadership that is aware of the changes that occur throughout the business, anticipates upcoming changes, and plans to optimize the new business system. In any case, they need to gather information about the elements of the business processes, analyze it, and then act.

In most businesses, the leaders traditionally focus on sales growth. They develop strategic plans to grow sales, estimate performance, and guess at the resulting profits. Under this system, success occurs by chance and profitability is suboptimal. This management by objectives implies managing the numerical objectives that have been set according to the strategic plan and manipulating those numbers to achieve the desired objectives. The problem is, however, that these kinds of strategic plans do not necessarily correlate to the business processes that contribute to the planned sales growth.

CURRENT ACCOUNTING SYSTEMS

Accounting systems play a key role in most businesses. Leaders implement great accounting systems to track every penny in the company. They hold meetings to assess the accuracy of the system, and big accounting firms are hired to audit the integrity of the accounting system, as well as to ensure compliance to legal requirements. Accounting itself has become such a big business that we need a new industry to audit the accounting firms—now there is an opportunity! With so much emphasis on accounting for pennies, it might seem surprising that there are dollars to be found in wastebaskets all around the company. Such waste is often readily visible to employees, yet ignored by the leadership. It is part of the hidden costs that go unaccounted for because of undue dependence on an accounting system that is incapable of tracking loss of business, wasted time, inefficiencies of business processes, and other hidden costs.

In other words, the traditional main source of information about business performance has been the accounting system, with leadership leading the business through financial measures. Such reporting systems can work well in a simple and monolithic business environment. However, when a business becomes more complex by diversifying its products or customer base or by increasing the number of processes or employees in the business, then supporting performance measures become not just helpful, but necessary.

INFORMATION AGE PARADIGM

Peter Drucker identified the need for new tools in managing business performance in the information age today.*According

*Harvard Business Review on Measuring Corporate Performance, Harvard Business School Press, Boston, 1998.

to his "new" paradigm, a business is a generator of resources, a link in an economic chain, an organ of society, and a creator of the material environment. He lists the following tools that generate information for executives in order to manage the business:

Activity-Based Costing (ABC).   To ensure the best return on investment, the need for each process is questioned, and the total cost must be managed. ABC shows the impact of each activity on the total cost.

Costs of the Entire Process Chain.   Because it emphasizes cost-based pricing in relation to pricing-based costing and other related processes, this tool requires a close relationship with suppliers, including sharing information and sometimes using the same accounting standards.

Information for Wealth Creation.   Most common accounting systems track costs instead of wealth creation. For executives to make informed decisions, they need four types of information:

1. *Foundation information.*   These are measurements such as receivables outstanding more than 6 months, cash flow and liquidity projections, total receivables, and sales.
2. *Productivity information.*   This includes the economic value of all costs, plus the cost of capital, production, and service workers, and processes in the value stream.
3. *Core competence information.*   This is the information needed to identify a business's core competencies as well as its competency for innovation.
4. *Resource allocation information.*   This includes conventional measurements, such as return on investment, payback period, cash flow, discounted present value, ratio of opportunity, and risks, to allocate capital or human resources.

Outside Information.   Because the environment outside a business affects its performance, information about markets, customers, technology, taxes, social legislation, distribution channels, and intellectual property rights needs to be organized. Accordingly, organization is a cost center, while the outside environment is the profit center.

Drucker implies that as circumstances change, innovative measurements are needed to monitor business performance.

SIPOC ANALYSIS FOR CONSTRAINTS

Business circumstances have changed greatly since Drucker first described his system. With the commercialization of the Internet, implementation of quality management systems such as ISO 9000, and shrinking profit margins, a business can be seen as a system in itself. Remember, a business is really a collection of processes, such as production, service, quality, purchasing, sales, and management processes. Just as any process needs to be optimized for best performance, the business process itself needs to be optimized for profitability.

One effective Six Sigma tool for analyzing the business process is *SIPOC* (*Suppliers, Inputs, Process, Outputs,* and *Customers*), as seen in Figure 4-1. There may be several suppliers (up to several thousand), inputs (up to several thousand), processes (up to several dozen), outputs (up to many products or services, and hundreds of related characteristics), and customers (several groups).

The SIPOC analysis shows that a business can have hundreds of constraints. Each constraint can contribute either to improving profitability or to creating waste, adversely affecting profitability. The challenge for management is to ensure that all these many variables or activities perform at an acceptable level. To fine-tune the performance of the processes, businesses need to create and monitor measurements that will illustrate the impact on profitability.

Historically, organizations have excellent systems for monitoring financial results; however, they often lack good monitoring of internal processes, including the leadership process. When the profit margins are acceptable and the business is profitable, the need for internal process measurement is easily

Suppliers	Inputs	Process	Output	Customers
Material suppliers	100% acceptable material	Leadership and management processes (communication, auditing, management reviews)	Product	Business customers
Information suppliers	Accurate information	Data analysis and reporting	Services	Consumers
Tools and equipment suppliers	Usable and well-maintained tools	Production/service delivery processes	Customer care	Stakeholders
Human resources providers	Skilled and available employees	Recognition	Value	
Training service providers	Sufficient capital	Benefits management		
Investors/shareholders	Continued interest in company	Training		
Office supplies suppliers	Office suppliers	Purchasing		
Freight services providers	Noninventory items	Marketing		
More...	Freight service and supplies	Sales		
	More...	Customer Service Documentation management		
		Calibration process More...		

FIGURE 4-1.   An example of SIPOC analysis for a business entity.

overlooked. But when one or more processes perform below an acceptable level because of external or internal challenges, profitability suffers.

When profitability suffers, leadership first considers quick fixes. Allocating limited resources for creating long-term solutions can be challenging or difficult to justify at this point. But the short-term mind-set can further erode profitability, and thus the downward spiral begins.

Cisco, after acquiring so many smaller companies, has almost real-time information about the financial performance of the entire company, so it can plan timely corrective action. Similarly, leadership must establish real-time measurements for the internal processes—a feedback system and corrective action to remedy unacceptable performance levels. In other words, the leadership must see the business as a large process that is supposed to generate profits for growth, better service

FIGURE 4-2.   Profitability management process.

or newer products for its customers, employee growth, and value to shareholders. (See Figure 4-2.)

MANAGING THE PROFITABILITY PROCESS

Managing the profitability process involves looking at inputs, outputs, execution, results, and the review process. The weakest link in managing the profitability process tends to be the review of business performance. In many businesses, management examines profitability much more frequently than it reviews the input streams of resources that will generate the profits.

When implementing a business performance measurement system, management must not forget to take into account the external factors that can rapidly hamper a company's performance. Although external factors are usually out of anyone's control, leadership must nevertheless monitor external factors and be prepared to handle them as effectively as possible.

For example, equipment manufacturers in the telecommunication industry were highly valued in the recent past. The industry itself was profitable and growing rapidly. Within a few years, however, many telecommunication businesses crashed. External factors, such as industry saturation, the bust of dotcom businesses, federal regulations, a new business model of service providers, and competition, led to overcapacity and the decline in industrywide performance. In such an environment, innovation by one company could immediately affect another company's sales and revenues adversely.

When Motorola launched the cell phone in the middle to late 1980s, it created a new industry. Motorola grew and its competitors struggled. After several successful years, however, Nokia's innovative products and a slack in Motorola's performance resulted in Motorola's market share declining while Nokia's growth exploded. This result, nevertheless, is a natural process, as competitors continually improve products and services. Customers and consumers demand better, faster, and less expensive products. The cycle of improvement moves from one company to another company. Motorola recognized that in the early 1980s when its leadership committed the company to dramatic improvement using the Six Sigma methodology. Today, Motorola is reemphasizing Six Sigma to improve performance and regain lost market share.

EXTERNAL FACTORS

The external environment requires businesses to perform competitive benchmarking, reduce costs through improvement and innovation, monitor customers' behaviors and attitudes, adjust to government regulations, improve response time through streamlining operations or outsourcing, and have accountable leadership. The desired leadership role has changed from strong management to passionate leadership. A business's leadership must be committed to the company for more than personal gain. Personal gains will only be realized over a sustained period through positive behaviors, exemplary leadership, and accountability to employees and shareholders. The recent debacles of large corporations imploding, as well as the disappearance of small businesses, have magnified the importance of responsibility and accountability within the leadership function. Leaders can no longer be simply charismatic and person-oriented; instead, they must embrace a sound process designed to achieve business objectives for the various stakeholders.

A good performance measurement system, then, explicitly considers not only external environmental factors and measurements for the rate of improvement, but also the role of leadership. What is valued must be measured. In today's business environment, where competition is increasing at an ever-rapid rate and where profitability must be guided as a missile, businesses need a scorecard that accounts for such factors.

PROCESS-BASED MEASUREMENTS

Alternative measurement systems have been difficult to implement because they are oriented to the leadership or executive level, not the execution level. For example, the Balanced Scorecard, which has been defined as a strategic management system, has been successfully implemented at only a small number of companies, compared to the number of companies that have implemented a quality management system. What's needed is a system that is strategic in intent, yet has very clear links to an organization's processes. Such a system must be easy and inexpensive to implement. Its measurement system must be easy to understand and must include analysis, reporting methods, and improvement mechanisms. Certainly in this information age, implementing such a system for quick summaries and reports is much easier than it was several years ago.

SIX SIGMA BUSINESS SCORECARD

The Six Sigma Business Scorecard is a complete corporate performance system that requires leadership to inspire, managers to improve, and employees to innovate to achieve the optimum level of profitability and growth. Based on my experience with more than 100 companies (small to large), I developed it through analyzing these companies' internal operations, participating in strategic planning at several companies, meeting with CEOs and staff, participating in the direct improvement of several operations, and being involved with Six Sigma at its conceptual stage by working with the late Bill Smith. The Six Sigma Business Scorecard incorporates proven business improvement practices. It introduces process measurements in a business irrespective of responsibility and authority, it pushes for a dramatic rate of improvement, and it holds the leaders accountable for business success through their passion and active involvement.

The Six Sigma Business Scorecard can be applied to businesses that offer products or services. It works for organizations that are private or public, small or large, government or education institutions, and with one or multiple locations. By viewing each business as a collection of processes, the management of each business process becomes an objective that is monitored using performance measurements. The Six Sigma Business Scorecard addresses the need for leaders to put their arms around the entire business and not just its parts. It allows them to be total in nature instead of fractional (strategic or operational), and so practically guarantees improvement instead of leaving them guessing at what will be successful. The Six Sigma Business Scorecard is the next step in the natural evolution of performance measurement systems.

BUSINESS PROCESSES TO CONSIDER

Typical business processes are shown in Figure 4-3. Within each process, there are subprocesses and activities where excellence must be achieved through effective measurements and monitoring. The Six Sigma Business Scorecard encompasses the following aspects of a business:

1. Purpose of business
2. Future performance

Business Processes	
Accounting and Measurements	Business Policies and Procedures
Learning and Innovation	Business Planning and Execution
Research and Development	Global Marketing and Sales
Administrative and Facility Management	Recruitment and Development
Maintenance and Calibration	Audits and Improvement
Leadership and Profitability	Production and Service
Information Management and Analysis	Partnerships and Alliances
Purchasing and Supply Chain Management	

FIGURE 4-3. Business processes.

3. All business processes

4. Measurements that can be aggregated to the corporate level

ELEMENTS OF THE SIX SIGMA BUSINESS SCORECARD

Considering the business's purpose, its processes, and its total culture, the Six Sigma Business Scorecard combines various measurements into seven elements, as shown in Figure 4.4:

1. Leadership and profitability

2. Management and improvement

3. Employees and innovation

4. Purchasing and supplier management

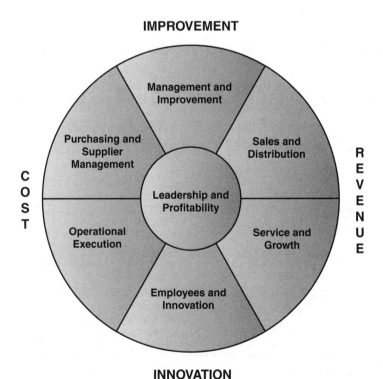

FIGURE 4-4. Six Sigma Business Scorecard.

5. Operational execution

6. Sales and distribution

7. Service and growth

These seven elements and their related measurements are shown in Figure 4-5. By no means is this set of measurements universal. The intent of each measurement is to assess effectiveness of activities critical to an organization's wellness and profitability.

Categories	Objectives	Measurements
Leadership and Profitability (LNP)	Lead company to wellness and profitability	Communication
		Inspiration
		Planning Accuracy
		Community Perception
		Employee Perception
		Employees' Recognition
		Compensation/Profitability
		Asset Utilization
		Return on Investment
		Debt-to-Equity Ratio
		Profitability
		Shareholders' Value Growth
Management and Improvement (MAI)	Drive dramatic improvement	Goal Setting
		Rate of Improvement
		Planning for Improvement
Employees and Innovation (EAI)	Involve employees intellectually	Employee Innovative Recommendations per Employee
		Investment per Employee
		Number of Patents or Publications per Employee
Purchasing and Supplier Management (PSM)	Reduce cost of goods or service	Material Acceptance
		Total Spend/Sales
		Suppliers' Defect Rate (Sigma)
		Suppliers' Involvement in Development
		Cost of Goods/Service Sold
Operational Execution (OPE)	Achieve performance excellence	Operational Cycle Time
		Process Defect Rate (Sigma)
		Customer Defects/Total Defects
Sales and Distribution (SND)	Manage customer relationships and generate revenue	Number of Inquiries
		New Business ($)/Total Sales ($)
		Profit Margins ($)/Sales ($)
Service and Growth (SAG)	Gain competitive advantage and grow	Customer Satisfaction
		Customer Retention
		Repeat Business ($)/Total Sales ($)
		New Product or Services
		Patents or Trademark

FIGURE 4-5.   Six Sigma Business Scorecard measurements.

1. LEADERSHIP AND PROFITABILITY

Any organization is only as good as its leadership. Leadership requires a strong sense of righteousness, a vision for setting the direction, values to make effective decisions at a personal level, and beliefs that define actions consistent with those values. The main characteristics of a good leader include honesty, knowledge, integrity, and a futuristic outlook. Good leaders look at their businesses as a continuation of their lives, with the purpose of achieving service to society, value to customers, and growth of their employees and themselves.

Leaders create the corporate culture based on the business's values and beliefs. True leaders do not ask employees to satisfy customers, deliver excellence, and do more while the leaders themselves ignore quality and excellence, demonstrate a careless attitude about customer needs, and spend a lot of time on the golf course. Such a poor display of care for the company and its customers promotes similar behaviors in employees. Leaders have additional responsibility, commensurate with their compensation, to make the right decisions. The impact of a leader's decision is multiplied by the number of employees in the business and the suppliers. If the leadership, in order to meet end-of-the-month revenue goals, takes one shortcut, the company's employees may likely make similar decisions throughout the month, triggering a downward trend in performance.

One good way to check consistency of actions is to examine the ratio of exceptions to standard procedures as well as the level of leadership involvement in daily decisions. Less leadership involvement in daily activities is preferred so that employees can then take an appropriate leadership role in their own areas. Leaders must make decisions at the corporate level and let employees make decisions at the process level. This allows employees to prepare for growth and innovation.

Another critical role of leadership is to maintain the financial health of the company. In no uncertain terms, the leadership is supposed to be accountable for the financial success of the company. For as long as businesses have existed, they have been using financial measures. As many as 50 of these financial measures can be used in conjunction with the process-level measurements. The financial success of the business correlates directly to its operational excellence. Good operational performance correlates directly to good financial results. Conversely, poor operational measures eventually lead to poor financial results. This relationship is so direct and simple, but it is still overlooked by management, as seen in many decisions on a daily basis.

Employees are almost always aware of this managerial oversight. They wonder why the leadership overlooks the facts of what employees see on the floor. Especially in tough financial times, this leads to employees' perception of leadership's hypocrisy and lack of commitment to corporate success. Employees believe that one bad leadership decision can have an adverse impact on hundreds of things that the employees do well.

Assuming that a company's leaders are well meaning, their seemingly inconsistent decisions can best be viewed as stemming from the absence of information. Any leader's first priority, therefore, must be to get current and accurate information about the state of business in order to make effective decisions. The Six Sigma Business Scorecard contains leadership measurements designed to enhance the role of leadership by providing leaders with subjective as well as objective feedback and to facilitate leaders in making decisions based on facts and experience rather than gut feelings.

We commonly read about examples of poor business leadership in the newspapers—after all, after-the-fact poor performance is what makes news. Examples of good leadership are more often hidden from the public. Good leaders have to publicize themselves through their actions. In the late 1980s when the guidelines for the Malcolm Baldrige National Quality Award were published, they reflected the behaviors of Robert Galvin, then CEO and chairman of Motorola. Those guidelines included performance expectations in terms of interaction with the community, positive behaviors, and the well-being of corporations.

No leader is a perfect person; like everyone else, leaders undergo a lifelong learning process. That's why performance measurements must promote such openness and leadership growth. Important leadership processes include the following:

1. Establish, communicate, and practice corporate values and beliefs.
2. Facilitate a strategic plan for the company.
3. Inject positive energy into the organization.
4. Be informed in real time through performance measures.
5. Commit to the well-being of employees, customers, and society.
6. Guide the staff to achieve the planned performance levels.
7. Challenge everyone to excel and innovate.

2. MANAGEMENT AND IMPROVEMENT

Management, so-called *middle management,* is a link between strategy and results. Conventionally, *management by objectives* (MBO) has been the mantra for *the* way of doing business. The MBO approach led to a system in which managers focus heavily on narrowly defined results. The objective becomes to meet or exceed numerical goals in one way or another. Ultimately, MBO has become a manipulation of numbers at all company levels—from the president to the operator.

This approach often leads to situations where process problems are overlooked. Because numbers can be easily manipulated without making the effort to solve a problem, problems often continue unchecked for years. Managers, talking about their daily activities, often complain, "My plate is full," or "I have a million things to do. I have so many fires to fight, I do not even know where to start." This very mind-set on the part of managers is what leads to the high cost of poor quality, which many consultants estimate to be as much as 40 percent of sales.

The main responsibility of the managers is to improve their departments or areas as planned and to produce expected results, not to manage the numbers and do whatever is necessary. The MBO approach must change to *MPO* (*Manage by Process Objectives*). Accordingly, managers must be responsible for understanding operations, setting expectations for employees to establish optimal processes, and ensuring expected results. This requires statistical thinking and an understanding that if the process is managed correctly, the expected results should occur. If the results are not as expected, the process needs to be adjusted.

As a process example, a manager told the engineer to produce electronics boards as fast as possible. The engineer bought a machine for soldering and set it at the highest temperature and belt speed to produce the highest number of soldered boards in the shortest amount of time. On paper it appeared that capacity goals were achieved. However, the high temperature of the machine caused some parts on the electronics boards to melt. To correct the melting, the manager complained to the design engineer. The design engineer designed a heat sink to absorb the extra heat. Additional heat was absorbed, changing the thermal profile of the board and causing some connections to become unsoldered and displaced. The result was a high defect rate.

The manager directed the process engineer to fix the defects, and the process engineer recommended setting up a repair and rework station. The number of good boards increased. When these boards were tested, yields were better although some defects were still found. The field performance of the boards, however, had degraded as more products failed in the field. The warranty cost increased.

Now, a customer service person or the manager goes to observe the failures in the field. The costs of these visits quickly add up. As a result, the profits from the entire operation suffer. The general manager, who does not know why profitability has decreased, establishes a quality improvement program, hires a consultant, and so on. Employees are trained in problem solving, but most of them didn't get much out of it because they could not understand why they were being trained—they currently had no problem. The program eventually leads to insignificant gains.

Similar examples exist in almost every company, whether service or manufacturing, software or hardware, or private or government. The point here is that profitability is a small number compared to total sales, and just one process can eat away the profitability of the entire operation. Just as one punctured wheel disables a car, a poorly managed process can kill a company. In several companies I have visited, a key individual at a senior engineering level or a manager level (well recognized and respected in the company) had been unknowingly hurting everyone else. In the absence of proper performance measurements, it is difficult to eradicate the situation, and the entire business suffers until its close.

Therefore, management must start the optimal process right away. They need to define and optimize each process, establish performance measurements, document the process, train the people, and capture and analyze performance data. If the results are not satisfactory, remedial action must be taken with a sense of urgency. When processes are managed in such a way, managers will not have a million things on their plates, and they will not be working late and forcing their people to work late to do tasks that do not add value. Management must keep the processes simple, measurable, and managed.

Many times, management personnel cannot relate the performance of their processes to profitability. In such cases, processes must be managed well; otherwise, they are guaranteed to hurt profitability. Every waste in the value stream is deducted directly from profitability. One percent waste in one process can account for as much as 25 percent of the profitability. On the other hand, it is possible for a business's sales to grow while its profitability is shrinking. This is often considered acceptable. No matter how big a business is, however, not one can last long by losing money or having waste.

Management's job is to enable its people to do well at work, accomplish a lot, and have fun doing it. However, due to increasing competition and pressures to reduce price and improve quality, management must also focus on how fast the process is improving. Even though no one has asked about process improvement speed, that concern must be the new mantra.

The performance level is less critical than the rate of improvement. Moreover, improvement of 5 or 10 percent is not enough, as no one actually sees that amount of improvement. If, however, the performance improvement is planned at 30 to 60 percent annually, everyone will have to be involved, innovation is required, and the improvement exceeds the measurement system errors. Improvement will be noticeably significant because everyone has participated in bringing it about. In the case of 5 to 10 percent improvement, management may know what to do, but no one else is aware of it or involved. The Six Sigma methodology requires an aggressive rate of improvement in the customer-critical areas.

3. EMPLOYEES AND INNOVATION

Not many businesses exist without employees. When no mistakes are made, employees are thought to be an essential and good component of all business processes. However, when mistakes are made, employees are almost always fingered for blame. In the past, employees were hired for their physical effort—to push the product through operations either to the stockroom or for shipping.

In today's workplace, with increased automation, computer networks, and other productivity tools, businesses need more informed and alert employees. Normally, in a business, managers believe they know a lot and make most of the decisions. In a growing organization, however, employees' intellect must be utilized as well. Employees must be challenged to think and perform better. Employees must also feel that they are contributing in a real sense, not just by doing what someone else tells them to do with no room for creativity.

A statistical distribution of employees' performance occurs naturally. In other words, most people do the necessary amount of work, while some are superior and a few are substandard. The main objective of a company's management is to maximize the contribution of employees in terms of physical and intellectual productivity—to bring out the best on the job every day at every job. This requires effective job planning, incentives for extra effort, recognition for prevention of problems, and reward for innovation.

The process for employee recruitment and development must be improved and positively offered to employees. A job must be an opportunity both to learn and earn. Surveys have always found that people like their jobs when they accomplish something in them. Sustained accomplishments are produced through continual learning. Therefore, cross-functional training, skills development, and empowerment must be provided to employees.

Contrary to the general understanding that innovation involves designing expensive products, innovation can occur at every process and by every employee. Given the right circumstances, every employee has a potential to become an inventor. Many times, innovation is born out of production employees, who are immersed in their jobs 8 hours or more per day. As Bob Galvin said, it is important to use the brains of all employees—not just those of managers. He recognized the potential of human beings.

Many people working in the production area are leaders in their community or experts in other areas. On the other hand, many of the managers recognized at work for their performance are intellectually inactive outside work. Innovation is a discovery process that requires broad experience, a purpose for applying cross-discipline experience, and encouragement for new ideas. Companies encourage innovation at different stages, such as having a suggestion box, developing a solution at work to improve a process or product, applying for a technical or engineering award, and applying for or receiving a patent. Companies encourage innovation by rewarding employees for publishing articles in magazines, presenting at conferences, speaking at professional associations, participating in standards bodies, teaching at local colleges, recognizing innovation in front of corporate executives, and even sharing gains or savings.

Successful companies, both small and large, encourage employees' total participation at work to bring out their best. Many small companies enlist employees' ideas and do well, while many larger companies, due to their bureaucracy, stifle employees' creativity and lose business. In future organizations, better, faster, and less expensive solutions and an aggressive rate of improvement will be realities that can only be realized through the intellectual involvement of employees. The learning organization will have a strong competitive advantage in all aspects of business. Such organizations will actually achieve better profit margins, because new products or services command a premium price in the marketplace.

4. Purchasing and Supplier Management

In manufacturing, about one-third of the cost of providing products or services is attributed to the costs of materials. In service industries, person-hours are a significant cost factor. Every company has a purchasing process and works with suppliers. Companies are always trying to reduce the cost of purchases. Large corporations have annual price reduction targets with which smaller suppliers have to work. The result? These large corporations are promoting the cheapest products or services instead of innovative and good products or services.

In the supply chain, a business must look into the total cost of a purchase instead of just the initial purchase price. This current approach to buy the cheapest products has caused many suppliers to take shortcuts in operations. Such shortcuts generally lead to a higher defect rate for their customers and increased paperwork to correct the problem.

Many industries relocate globally in search of cheap labor. The whole mind-set of cheap labor, however, does not necessarily bring value, whether in the area of hardware or software. Companies seeking such cheap labor often overlook other costs incurred by challenging geopolitical issues in developing nations.

To optimize the purchasing process and the associated costs, relationships with key suppliers must be managed well. Managing suppliers' performance includes understanding the suppliers' capability, resources, ability to rapidly adjust to changing demands, willingness to innovate during product development, and commitment to an aggressive rate of improvement. With current outsourcing trends and increased dependency on suppliers, a company's capability is only as good as that of its weakest supplier. One weak supplier somewhere can become a bottleneck in the entire supply chain.

Businesses who are themselves large customers must view their suppliers as extensions of their operations and continue to be actively involved in the improvement process. The supplier management processes are defined by the quality management systems a company has implemented. For those systems to have value, the company must establish clear expectations with its suppliers in terms of technical, business, and service requirements. Supplier performance can then be measured against clearly defined goals as well as the rate of improvement.

5. Operational Execution

Most companies have difficult times implementing a sound strategy effectively throughout the corporation. Strategic blunders throughout a corporation can lead to its major fall. However, the main reason for mediocre performance and profitability is poor execution of simple, well-known principles. Many companies devote resources to developing a sound corporate strategy without somehow recognizing the need to execute it well. They assume people will just figure that part out. Without a well-thought-out execution and institutionalization process, however, the new ideas, strategies, or programs become the cannonball of the year. Employees become immune to new initiatives about which the management is so excited. Leadership thinks it has assigned people to tasks and empowered them to perform; therefore, it now blithely expects results.

When a major initiative is launched in an organization, the leadership (CEO or equivalent) must be passionately committed to it. Unfortunately, the whole process of commitment is often reduced to a few slogans. The initiative simply stays a slogan without producing any results. Commitment must instead be based on the real understanding by the leadership of its potential benefits at both personal and corporate levels. Such an approach allows employees to see something for themselves and for the company. If the potential return on investment for their effort is clear and visible, they are interested in working on it; otherwise, no matter how bright an idea or how strong the leadership, the strategy will not yield the desired results.

In the case of Six Sigma initiatives, many companies have started to train people in herds, expecting miracles. But the miracles are not happening. I was at a company recently where I learned about the performance of its internal processes and was amazed that the company even survived as long as it had. Two days later I read in the newspaper that the company had saved hundreds of million dollars through its Six Sigma initiative. If I had asked the employees about the improvement, however, they would have said that they had not seen a penny from all the savings. To the employees, the savings appeared to exist because of the management's number manipulation techniques rather than because of any actual improvement.

Each Six Sigma project at a large corporation is expected to save about $100,000 to $250,000. Some of these large corporations, with the help of recognized consulting firms, have initiated thousands of projects. Recent feedback from one such company is that about 10 percent of these projects are producing significant savings. At another company, a CEO had committed to Six Sigma, hired a consulting company, trained all his professionals (more than a thousand), and initiated many more changes to transform the culture. Those trained Black Belts, however, are having a difficult time identifying projects for improvement. Eventually, because the strategy was poorly executed, many of these employees will leave the company for a better work environment.

Larry Bossidy and Ram Charan, in their recent book *Execution* (2002), have identified the building blocks of execution. Those building blocks include knowledge of people and the business, realism, clear goals and priorities, follow-through, rewarding the doers, employee growth, and self-awareness. What I have seen in many corporations is that execution fails because someone failed to plan to do the job well. The implication here is that the resource requirements needed to carry out the plan well are quite different from those needed to just do it. Without visualizing the end product or service, people fail to recognize how the project will be completed successfully and what will be the measures of successful completion.

Another flaw lies in understanding empowerment. Empowerment includes education, authorization and accountability, and recognition. Businesses poorly implement all three aspects of empowerment. For example, we establish the expectations of employees before signing up for training. We also give employees time off from work life by sending them to a seminar.

Training employees and expanding their skills are one of the best investments a company can make, especially when the employee can use the skills learned in the training class. There is also value in simple awareness, however; that is a short-term benefit with no significant impact on the bottom line.

The authorizing aspect of empowerment means communicating an assignment and then forgetting it. Authorization includes clearly communicating the expectations of the project, the expected behaviors and processes, and the accountability for results. Just as there is a reward for excellence, there must also be a lesson learned from failures, which are not to be repeated.

In many companies, people do the work and try their best while all the while not really knowing what is expected of them. In other words, they really do not understand what and how they are supposed to perform at work. Lack of clearly defined expectations, responsibility, and accountability leads to suboptimal performance instead of peak performance. Among these employees indecisiveness prevails and dependence on superiors increases. Through better checks and balances, empowerment occurs in a subtle manner to achieve common objectives.

Change is routine in any business. There are no two identical days, strategies, or even people in this world. Everyone must be prepared to handle change. For a group in a business, the change process must be managed based on sound building blocks.

John Kotter, in the book *Leading Change* (1996), has identified eight stages of implementing a change. The first stage is to establish a sense of urgency. We allow too much complacency in the early stages of the project, assuming there is still a lot of time to complete it. Producing initial successes, like small wins, is very critical in order to understand the details of the new strategy and optimize the execution process. The initial phases of the change process are overlooked. We end up having a change process that lacks urgency, doesn't provide mentoring, and poorly communicates expectations.

Another stumbling block is the decision-making process. Uncertainty about doing a task is worse than doing it wrong or not doing it at all. Uncertainty creates confusion and affects others. Indecisiveness comes from a lack of information and a fear of failure. The leadership must reward innovation and risk taking, and it must also establish a culture to make decisions and move on them. That creates a dynamic and ever-changing organization, as people are willing to try different things and do not fear failure.

Well-executed strategy can be measured in terms of its performance against expectations and its timeliness. If we do a great job when the customer does not need it, that is just as bad as performing a job poorly. Every requirement has a timeliness component associated with it, and it must be given equal significance. The challenge is to define expected performance levels and establish appropriate timeliness requirements associated with them. Performance can be measured in terms of C_p, a ratio of expected to actual performance levels:

$$C_p = \frac{\text{Expected performance level}}{\text{Actual performance level}}$$

In manufacturing operations, the expected performance level is predefined tolerance, and the actual performance level is the process capability in terms of its variance.

Another measurement, C_{pk}, measures the shift from the target, or closeness to the limits, to estimate the reject rate. In general, C_{pk} can be stated as follows:

$$C_{pk} = \frac{\text{Nearest limit} - \text{Average performance}}{\text{Inconsistency in process}}$$

In other words, we need to measure how much inconsistency we have in our processes compared to expected inconsistency, and how far we are from the established target performance level. These are statistical measures and are a good set of performance measurements. The total cycle time and the attempt to reduce the total cycle time enable us to identify waste of resources and streamline processes.

6. SALES AND DISTRIBUTION

Just like purchasing, sales and distribution are well-understood processes. And again, the emphasis should be on the sales of value—not on the cheapest product or service. For a company to be profitable, certain margins must be maintained. In some companies, for example, sales representatives receive a commission from every sale, while the companies themselves lose money in delivering the promised products and services.

The sales effort consists of identifying new business by building relationships. The intent must always be profitable growth. I have seen companies trying to be billion-dollar companies within a year that are starting from the ground level. I have also seen companies where new business accounts for only 5 percent of the total business, and they depend heavily on a customer or two for their success. For a company to be successful, there must be a balance between total sales and the ratio of new business.

Since the success of the sales process is viewed as getting new business, that is how it must be measured. The network of dealers and distributors selling your products and services must be similarly managed as a process with clear expectations for sales and margins. When a business exists just to keep people busy, it achieves exactly that result—keeping people busy—but no profits. Businesses do not last long this way.

7. SERVICE AND GROWTH

Superior customer service, in conjunction with superior products or service, is critical to growing the business. The purpose of superior customer service must be not only to understand customer concerns but also to anticipate customer requirements. Customers have implicit and explicit requirements, subjective and objective expectations. They love to be cared for by their suppliers.

Customer service can provide the insight into customers' unspoken requirements (features or service customers want to receive) that they would love to have met. Segments of the automotive industry, for example, have grown when they succeeded with superior customer service. Service has become integral as the relationship between customers and suppliers becomes stronger and more dependable. Customer service can identify opportunities—from design to production—for suppliers to fulfill.

Meeting only the minimum requirements makes suppliers vulnerable to being yanked by the customer at any time. Customers are rarely loyal. Even doing a good job in meeting customers' spoken requirements does not guarantee success. Superior service, however, bonding between the customer and the supplier's employees, and a better understanding of each other's needs facilitate a mutually beneficial relationship that both parties nurture as a true partnership. Sales can bring in a new customer, and customer service can build that relationship to get repeat and additional business. Sales works to acquire new customers; customer service retains them.

SIX SIGMA BUSINESS SCORECARD MEASUREMENTS

With the seven elements or categories of the Six Sigma Business Scorecard defined, we need to look at the measurements associated with them. These measurements are oriented more to process than to function.

The challenge lies in deciding which measurements to choose and which to exclude. For example, suppose that in terms of financial measurements, 30 to 40 different measurements can be easily identified in terms of liquidity, efficiency, profitability, and so on. Similarly, there are dozens of potential measurements within internal operations. Therefore, the Six Sigma Business Scorecard has identified key process indicators to help establish a relative indicator of business performance.

Some measurements appear to be unconventional (i.e., they look unique or unsubstantiated). However, the intent of these measurements is quite clear. Key roles and related processes were identified when the categories were created in order to establish accountability as well as criteria for effectiveness. The measurements listed comprise a set of sample measurements applicable to most businesses. If a measurement is not applicable, it can be replaced with an alternative measurement or ignored altogether.

The main purposes of the measurements are to challenge the existing system and identify opportunities for improvement and profitability. There is no absolute system that can be used from company to company. Each company's culture and measurement system are different. The system is not prescriptive; local adaptation is almost mandatory. This is so because the final measurement system must reflect true indicators of variances in a company's performance.

BUSINESS PERFORMANCE INDEX

How does a company establish a corporate-level performance index that quickly delivers a snapshot of the business performance to the CEO or the executive staff? The Six Sigma Business Scorecard includes a *Business Performance Index* (BPIn) based on the 10 critical measurements that relate to the wellness of the company. It addresses various elements of the Six Sigma Business Scorecard and corresponding measurements (see Figure 4-6). BPIn is a sum of weighted corporate performance in various categories of the Six Sigma Business Scorecard.

Measurements	Category Abbreviation	Category Significance	Performance Against Plans	Index Contribution
1. # Employees Recognized for Excellence	LNP	15	50	7.5
2. Profitability	LNP	15	75	11.25
3. Rate of Improvement (All departments)	MAI	20	60	12
4. Recommendations per Employee	EAI	10	60	6
5. Total Spend / Sales	PSM	5	80	4
6. Suppliers' Defect Rate (Sigma)	PSM	5	60	3
7. Operational Cycle Time Variance from Planned	OPE	5	60	3
8. Process Defect Rate (Sigma)	OPE	5	80	4
9. New Business ($)/Total Sales($)	SND	10	90	9
10. Customer Satisfaction	SAG	10	80	8
Corporate Wellness (BPIn)				67.75%
Corporate DPU				0.3893
Corporate DPMO (15 executives)			25,956	
Corporate Sigma				3.44

FIGURE 4-6. Example of Business Performance Index.

The Measurements column identifies one or two key business indicators, and the Category Abbreviation column relates the measurement to the corresponding element of the Six Sigma Business Scorecard. The Category Significance column ranks the importance of each category. The Leadership and the Rate of Improvement categories, for example, are ranked highly significant because of their direct correlation with profitability. Leadership and profitability are weighed 30 percent, and the rate of management and improvement is ranked 20 percent. The Performance Against Plans column assesses a corporation's performance against its plans. For a company to rank well, it must have a sound plan, effective execution, and good results. The Index Contribution column includes the weighted contribution of each category determined by multiplying the Category Significance and Performance Against Plans values.

CORPORATE DPU AND DPMO

The Corporate DPU is a measure of opportunities for improvement that exist for improving profitability and growth. Corporate DPU is calculated by using the formula

$$DPU = -\ln \frac{BPIn}{100}$$

This DPU is then converted to a Corporate DPMO. Corporate staff size represents the complexity of an organization and is given credit for the organization's success. Accordingly, any major error made in the business must be the responsibility of one of the executives. The formula for calculating Corporate DPMO is as follows:

$$\frac{Corporate}{DPMO} = \frac{DPU \cdot 1,000,000}{Number\ of\ executives\ reporting\ to\ CEO/COO}$$

where the number of executives reporting to the CEO or COO represents opportunities to make mistakes in decision making.

CORPORATE SIGMA LEVEL

Once the Corporate DPMO is calculated, the corporate BPI is converted to a Sigma measure. Figure 4-7 shows Sigma values and related Corporate DPMO levels. Using this table, the Sigma level associated with the Corporate DPMO is determined.

COMPARING BALANCED SCORECARD AND SIX SIGMA BUSINESS SCORECARD

Figure 4-8 identifies differences between the Balanced Scorecard and the Six Sigma Business Scorecard. Because of trends in the business environment, market dynamics, the rapid evolution of information technology and the Internet, and competitive pressures, profitability is shrinking. However, businesses have significant opportunities that could improve profitability. Without an integrated measurement system that relates to the corporate culture, vision, and beliefs, identifying

Corporate DPMO	Sigma	Corporate DPMO	Sigma	Corporate DPMO	Sigma
691,462	1	115,070	2.7	1,866	4.4
655,422	1.1	96,800	2.8	1,350	4.5
617,911	1.2	80,757	2.9	968	4.6
579,260	1.3	66,807	3	686	4.7
539,828	1.4	54,799	3.1	483	4.8
500,000	1.5	44,565	3.2	337	4.9
460,172	1.6	35,930	3.3	233	5
420,740	1.7	28,717	3.4	159	5.1
382,088	1.8	22,750	3.5	108	5.2
344,578	1.9	17,865	3.6	72	5.3
308,537	2	13,904	3.7	48	5.4
274,253	2.1	10,724	3.8	32	5.5
241,964	2.2	8,198	3.9	21	5.6
211,856	2.3	6,210	4	13	5.7
184,060	2.4	4,661	4.1	9	5.8
158,655	2.5	3,467	4.2	5	5.9
135,666	2.6	2,555	4.3	3.4	6

FIGURE 4-7.   Corporate DPMO and Sigma level.

Balanced Scorecard	Six Sigma Business Scorecard
1. A strategic management system.	A performance measurement system.
2. Relates to a longer-term view of the business.	Can provide a snapshot of a business's performance, as well as identify measurements that would drive performance toward profitability.
3. Designed to develop a balanced set of measurements	Designed to identify a set of measurements that impact profitability.
4. Identifies measurements around vision and values.	Establishes accountability for leadership for wellness and profitability.
5. Critical management processes are to clarify vision/strategy, communicate, plan, set targets, align strategic initiatives, and enhance feedback and learning.	Includes all business processes, management and operational, i.e., leadership, innovation, rate of improvement, sales, service, purchasing, and production operations.
6. Balances customer and internal operations without a clearly defined leadership role.	Balances management and employees' roles; balances cost and revenue of heavy processes.
7. Emphasizes targets for each measurement.	Emphasizes aggressive rate of improvement for each measurement, irrespective of target.
8. Emphasizes learning of executives based on the feedback.	Emphasizes learning and innovation at all levels based on the process feedback. Enlists all employees' participation.
9. Focuses on growth.	Focuses on maximizing profitability.
10. Heavy on strategic intent.	Heavy on execution for profitability.
11. Management system consisting of measurements.	A measurement system based on process management.

FIGURE 4-8.   Balanced Scorecard and Six Sigma Business Scorecard comparison.

what contributes to the loss of profitability and growth can be daunting. Using lessons learned from the Six Sigma methodology, the Six Sigma Business Scorecard offers a method for maximizing corporate profitability. By integrating the existing quality management system with a performance measurement system, a corporation can aim for dramatic improvement in performance and significant improvement in profitability.

TAKE AWAY

1. There is an overemphasis on accounting while a lot of money is found in the wastebaskets all around the company. Such waste is readily visible to the employees, and is a "hidden cost" for management.

2. Executives need four types of information to make informed decisions—foundation, productivity, core competence, and resource allocation.

3. The organization is a cost center while the marketplace is the profit center.

4. Just as any process needs to be optimized for best performance, the business process itself needs to be optimized for sustained profitable growth.

5. The Business Scorecard is a complete and balanced corporate performance system that requires leadership to inspire, managers to improve, and employees to innovate for achieving sustained profitable growth.

6. Seven elements of the Business Scorecard include:
   i. Leadership and Profitability
   ii. Management and Improvement
   iii. Employee and Innovation
   iv. Purchasing and Supplier Management
   v. Operational Execution
   vi. Sales and Distribution
   vi. Service and Growth

7. Business Scorecard measurements are oriented more to processes than to functions or silos. The Business Scorecard aggregates into a Business Performance Index (BPIn) based on the ten critical measurements that relate to the wellness of the company.

8. BPIn is a number CEO must know in order to continually achieve sustained profitable growth.

9. BPIn is translated into a corporate DPU, a measure of opportunities for improvement that exist for improving profitability and growth.

10. Any error made in the business must be the responsibility of one of the executives reporting to CEO or COO.

PLANNING FOR THE SIX SIGMA BUSINESS SCORECARD

The Six Sigma Business Scorecard helps executive leadership and shareholders not only understand the company's performance through simple measurements, but also plan success. The Six Sigma Business Scorecard can help create a system that will improve the company's performance.

LEADERSHIP AND IMPROVEMENT

Many companies neglect to plan for improvement, in both good times and bad. When the economy is strong, margins are good, and the company is profitable, the leadership does not sense the urgency to improve profitability through strategic planning. On the other hand, when the economy weakens, margins shrink, and profit evaporates faster than steam, the focus is directed to fighting fires rather than planning. Rather than plan to collect meaningful operational data, leadership develops a plan backward from sales numbers and hypothetical profitability projections. The numbers are allocated to managers, who then create measurement methods to achieve the desired numbers. When the net results of this effort are not good, the real problem areas become even more difficult to identify. Thus begins the spiral of cost management: cutting expenses, reducing the workforce, and streamlining the product or service portfolio.

To be successful, leadership must gather information that accurately demonstrates what is happening in the business. Leadership must identify the key business measurements that indicate corporate wellness, gather operational performance data, identify opportunities for improvement, and use all this information to develop a strategic plan to improve business performance. The urgency to achieve the desired results must be clearly understood by the executive and management teams of the company.

EXTENT OF IMPROVEMENT

What extent of improvement is appropriate to aim for? When a business sets a goal to improve performance by 10 percent per year, employees are likely to complain that no one in the company feels any improvement. In fact, they may even feel that the company's performance has degraded. The reason is that the tangible results of about one-half of any improvement in performance may be consumed by cost-of-living adjustments. Another portion of the improvement may be attributed to measurement errors (whether intentional or not), and the remaining improvement may be attributed to real improvement in a few areas. As a result, most employees see no improvement, practically speaking. More significantly, the resulting improvement correlates insignificantly with improvement in profitability.

Suppose a company's CEO sets a goal to improve business performance by 10 percent. The management team is requested to submit a plan that will achieve the expected results. Naturally, they look for areas that can easily be tweaked. The 10 percent improvement is quickly realized, and everyone celebrates the success.

There's a better way to plan, however. Suppose that the CEO asserts that an aggressive rate of improvement is critical to maximize profitability and assigns challenging rate-of-improvement goals to all management team members. Practically everyone has the same goal: to improve at the specified rate. Typically, managers react with dismay, "This rate of improvement is impossible! We have never done this in the past." After the initial shock, however, when they come to understand the current problems and accept the common corporate goals, the goal of aggressive rate of improvement is accepted. Then the challenge to improve current processes begins. The way to achieve such aggressive rate-of-improvement goals is to examine and test the existing processes and look for a new way of doing them. To institutionalize and sustain such a high rate of improvement, the corporate culture must change. The cultural change begins with a new vision and new enthusiasm.

Some call this cultural change *corporate renewal*; others call it *reengineering the corporation*. Whatever it's called, the change must be driven by the new, clearly established corporate performance goals. Creating the vision is a critical step that must be led by the company CEO. Why should a leader be interested in achieving the corporate performance goals at an aggressive rate of improvement? Well, there must be something in it for the CEO personally, including financial incentives, in addition to the larger corporate goal.

BUSINESS OPPORTUNITY ANALYSIS

To achieve financial goals, a clear relationship must exist between the opportunities and the profitability. A business opportunity analysis will establish a baseline and identify the opportunities for improvement. If a company has done limited data collection up to this point, the Six Sigma Business Scorecard measurements can be used as a starting point. If an extensive data collection system already exists, however, those data can be analyzed and used to develop a business model that will identify what drives that company's profitability.

With a full understanding of the relationship between opportunities and profitability, being able to visualize the company's actual opportunities, anticipating trends in the industry, and taking into account plans for future growth, the company's leadership can commit to implementing the Six Sigma

Business Scorecard and improving the company's BPIn aggressively. The following relationships, although well understood, must be constantly monitored (see Figure 5-1):

Profits = Revenue − Cost of goods sold − Operational expenses

Growth = Function (Profits, External environment, Internal innovation)

Once a company's executive understands the potential for profits and its impact on compensation and value to shareholders, it is easier for that executive to commit to the strategic plan that facilitates it. He or she must be further aware, however, of external environmental factors (such as acquisition and mergers, competitive performance, government regulations, societal changes, and economic trends) in order to achieve desired growth.

ORGANIZATIONAL ADJUSTMENTS

A business must be organized as shown in Figure 5-2 to balance profitability and growth. The executive leadership team includes the *chief financial officer* (CFO), the *chief growth officer* (CGO), and the *chief operating officer* (COO). The CFO must ensure that the rate of improvement occurs as planned

FIGURE 5-1.   General profit and growth paradigm.

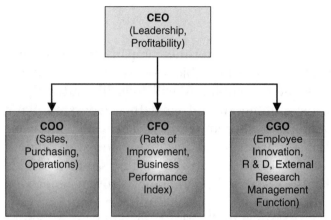

FIGURE 5-2.   General profit and growth functional layout.

as well as monitor the various business performance indicators that correlate to rate of improvement. The CGO, in the place of *chief technology officer* (CTO), must monitor external factors and employee innovation beyond the internal research and development activities. The COO is responsible for executing the operational plans within the organization as well as maintaining an eye on any variances in performance. Once the roles, responsibilities, and benefits of key executives are understood and defined along the lines of the Six Sigma Business Scorecard, their commitment starts to build.

Executive commitment for any initiative has become a buzzword at many companies. Such commitment is often demonstrated in letters or memos posted in the lobby and throughout the company. In many companies, these are commonly disregarded because employees observe the contradiction between the words and actions of the executive leadership. When the leadership is committed only on paper and not in deeds, employees will follow suit. The key to gaining CEO commitment, and hence that of the rest of the company, is through demonstrating the benefits of aggressive change: better performance, lower costs, higher growth and profitability, and superior financial incentives for executives and employees. The commitment begins with setting a Six Sigma Business Scorecard vision for the company.

VISION

To establish a meaningful vision for the company, the leadership team must consider its personal objectives for the company—in other words, what the company will look like after those objectives have been achieved and what the leadership plans to do for the company in the coming years. The leadership team should consider the impact of its actions on the stakeholders (i.e., employees, customers, suppliers, and shareholders). Leaders should also discuss the company's growth objectives and what resources and changes might be necessary to achieve that growth rate. Finally, the leadership team should consider the global competitive environment, required rate of improvement, and resources. All these factors taken together will provide a sense of urgency and identify the efforts required to achieve such goals.

Remember, the goals should not be easy to achieve. They must require significant effort and energy, and they must have the potential for a significant reward in terms of accomplishment, learning, money, and recognition. It is vital that the goals create excitement, nervousness, a sense of challenge, and the drive to do something differently. Executive leadership must take all these factors into account in creating a concise and compelling vision. The compulsion to achieve the vision must be visible not only in their words but also in their actions. This is so important that what the vision says is less important than how the vision is communicated. The executives' hearts must be immersed in achieving the vision, and that energy must be reflected by their body language, actions, attitudes, and decisions.

Although all corporate visions must lead to superior financial results, achieving superior financial results is not the only vision. The vision also reflects what a company will do to achieve better customer satisfaction and higher quality, increase market share, develop new products, or target new market segments. The vision relates to a specific need, which in turn relates to the company's core competency as an enabler to become a best-in-class or superior enterprise.

To achieve a company's vision, its beliefs or values determine the decisions executives and employees will make. If the company's core belief is "We must serve the customer to the best of our abilities," then decisions must be based on the impact on the customers' perception of the product or service. Our values reflect how our behaviors guide us in our daily activities. The values include respect for each other, honesty and integrity, or care for the company and its customers. Executive leadership plays an important role in establishing the guiding principles for the company. Successes and failures are part and parcel of any initiative; however, personal values are constant. The values, beliefs, and integrity inherent in the vision inspire employees to stay the course, demonstrate perseverance, and maintain personal commitment.

GOALS

Once the vision and values are aligned, leadership can begin to establish strategic goals. Strategic goals are the higher-level components of the vision that must be accomplished to produce the desired results.

Establishing strategic goals requires developing a business case, as well as understanding the cause-and-effect relationships between the vision and goals and the intended direction or trends. The business case begins with establishing the baseline performance level, the entitled performance level, and the desired performance level. The economic justification for achieving the vision and goals must be clearly understood. The goals then flow down into objectives and targets related to the various functions in the organization, the capability of those functions, and available resources. The management team then establishes the targets that will lead to the desired financial objectives (ultimately the profitability or money goals).

The success of an initiative depends upon what information is gathered to monitor the progress, as well as how it is gathered, analyzed, and acted upon. Management must design an information system to identify, analyze, and report the required information. The information system must clearly specify sources of input or data gathering as well as their aggregation for analysis.

CORE COMPETENCIES

Gary Hamel and C. K. Prahalad, in *Competing for the Future* (1994), state that the strategic architecture is a blueprint that shows how various functions should be deployed, what new competencies the company needs to acquire, or how existing competencies can be refigured to meet new goals. It helps align the organization to its planned initiatives, assigns leaders to each task, and outlines critical milestones. Without a high-level architecture to guide the process, detailed plans will lead to conflicting priorities and fragmented deployment of resources and efforts. This translates to a higher potential for failure. In many cases, detailed plans can even lead to the kind of false starts that have plagued many corporate initiatives, resulting in their abandonment and a change in leadership. When this happens, the entire organization is disrupted. Unwanted outcomes, even bankruptcies, have been the result.

SYSTEMS THINKING

Each organization, division, or department is an integrated and interrelated part of a larger system. Every part of the system affects the performance of other organizations in a larger system, such as in society or even the universe. Each level of a natural system interfaces or collides with other systems at the same level. Individuals work with other individuals, groups or departments interact with the whole organization, and the organization interrelates with the outside world. Managing these interactions is critical in an organization encompassing a new vision, if it is to lead to dramatic improvement and higher profitability.

Organizations naturally exist in a continuum where the environment, the organization, and feedback are interconnected.

According to Stephen Haines, the author of *The Manager's Pocket Guide to Systems Thinking and Learning* (1998), systems thinking involves thinking backward. Define first the vision, then the goals that will let the organization realize that vision, while simultaneously considering the environment as it will exist in the future. By establishing processes to realize vision and goals, we can then create measurement criteria that describe the specific factors defining success. These clear and consistent measurement criteria are used to evaluate the strategies' effectiveness.

EMPLOYEES' INVOLVEMENT

A few companies are able to execute business strategies effectively to meet their objectives. Linking strategy to the operation and operator levels requires employees to work effectively toward the objectives. Conventionally, businesses have assumed that the people at the working level contribute through their physical effort—their time and their productivity. Managers were assumed to be the only ones who should think and make decisions. However, no initiative can be even fractionally successful without the intellectual involvement of all employees.

In a recent Six Sigma Green Belt training class for a Fortune 50 company, participants said they had no incentives to cooperate with their leaders and Six Sigma Black Belts. They understood that the Black Belts are the chosen leaders and the Green Belts are there to assist them. If a project succeeds, Black Belts get stock options, recognition, promotion, and all the other benefits. The Green Belts, however, get no tangible rewards for the success of the project. When true teamwork does not exist, projects will not succeed.

In another company, a consultant was hired to improve the productivity of employees who had already been pushed to the limit. Employees told me that when the productivity consultant went there to work with them, they slowed down. They felt that the consultant had been imposed on them by the management to extract greater effort from employees and that neither the consultant nor management understood the employees' concerns. When productivity did not improve, the productivity consultant was fired, quality degraded, and the company closed. In this case, everyone on the floor knew the state of the business except the leadership. Employees knew that the leadership favored some employees over others and did not objectively try to achieve business goals. No plan was communicated to employees for achieving total customer satisfaction. Instead, the leadership always pushed employees to produce more while ignoring their ideas for improvement.

Surveys reveal that employees' loyalty improves when they accomplish a lot; that happens when employees are intellectually challenged and recognized for their accomplishments. To implement the Six Sigma Business Scorecard and an appropriate strategy to improve profitability, people must be challenged and recognized. To empower people, five things must be done:

1. Define roles and match people appropriately to those roles.
2. Delegate ownership of goals.
3. Establish a framework for performance and accountability.
4. Enable employees to develop or acquire necessary skills.
5. Recognize employees' efforts as well as results.

Employees must clearly relate the consequences of marginal performance and the rewards of excellence. Leadership must create the sense of urgency to achieve results. Good companies maintain the same level of urgency in good as well as tough economic times. Good work ethics motivate people to perform in a long-term way; incentives that are short-lived instigators do not.

TEAM STRUCTURE

A sound strategy can be executed effectively at all levels if the team charged with carrying it out is headed by a leader instead of a manager. The team formed to execute the strategy must have a few leadership players, several role players, and several doers.

A typical team, depicted in Figure 5-3, has three membership tiers. Level 0 establishes the expectations, level 1 provides the guidance and direction, and level 2 executes the strategy at the process level. The level 0 responsibility is assigned to a leadership-quality executive, the level 1 responsibility is assigned to a team consisting of key executives, and the level 2 responsibility is assigned to the process or department managers. The department managers then develop a series of measurements according to the Six Sigma Business Scorecard guidelines.

UNDERSTANDING MEASUREMENTS

Remember that business is a collection of processes. The processes might be office or production, sales or purchasing,

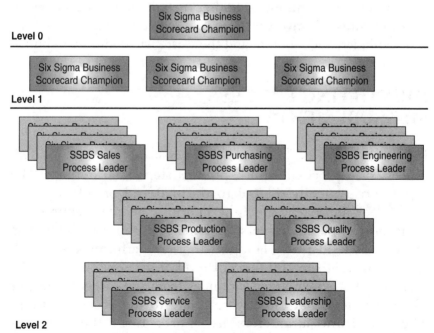

FIGURE 5-3.   Six Sigma Business Scorecard implementation team structure.

human resources or leadership. At any given time, not all processes are equally important for improving the business performance. Based on efficiency, quality, customer satisfaction, cycle time, and cost, some processes are prone to errors or waste while other processes perform as planned. The Six Sigma Business Scorecard requires understanding the Business Performance Index (BPIn) measurements in each process category and reviewing the operational effectiveness in that context. Performing a business opportunity analysis, to identify areas that adversely affect the profitability and growth, is a good way to ensure measurements have a specific purpose rather than just giving nice-to-know information.

Measurements are crucial tools used to identify opportunities for improvement, monitor progress, and inform the leadership about the state of the business. However, the most important reason for taking measurements is to make problems visible so that they can become opportunities for improvement. Good measurements identify or magnify the problem areas so they cannot remain hidden. If problems are visible, they cannot be ignored and thus will have to be fixed. Otherwise, they continue to bother the company's leaders as an area to address. Conversely, if problems are not identified, they will be hidden and ignored, and the waste will continue to build.

IDENTIFYING PROCESS MEASUREMENTS

Identifying process measurements begins by looking at all aspects of a process: inputs, process steps, and outputs. The inputs are typically related to the material or information, method or approach, machine or tools, people resources or skills, environmental factors, and measurement capability. The objective is to identify (1) inputs that are performing at a high level (almost at the 100 percent level of acceptability), (2) the variables that show larger variances from the target values, or (3) parameters that directly correlate to the process output. The significant parameters are chosen to be the measurements at the process level for monitoring.

Sometimes it is difficult to quantify measures for certain parameters, and it requires creativity to establish a realistic, practical, and economic measurement that will quantify variances in the process. The measurement method does not have to be perfect in terms of absolute measurements. The rate of improvement, instead of the absolute value, becomes more critical in making positive contributions to the process output and the ultimate profitability. For example, let's look at two competing car manufacturers. One has a starting defect rate of 130 defects per unit and the other, 110 defects per unit. At the end of 2 years, if both companies have achieved their planned rate of improvement of 10 times in 2 years, the defects per unit will be 13 and 11, respectively. In both cases, the results are much better than what they started with. Therefore, it makes sense to start the improvement process as fast as possible rather than wasting resources in excessively fine-tuning the measurement method.

When a measurement cannot be easily defined, the following questions help guide devising a meaningful measurement:

1. What is the intent of the process?
2. How do we know the process is producing good output?
3. Which process input or in-process parameters affect the process output directly?
4. Which process parameters are more critical than others?
5. How do we ensure that the critical parameters are operating at established values?
6. How will we collect and analyze the data to establish relationships between the process output and the input or in-process data?

Responses to these or similar questions will lead to the process measurements. Processes within a department may have multiple measurements. Establishing a defect rate at each process and aggregating that rate for all processes in each department can facilitate a measure of the department's performance. For each department, quality, Q, timeliness, T, and cost, C, play a role in establishing measurements to improve the department's performance.

Think about what measurements might be devised for a purchasing department, for example. The purpose of a purchasing department is to ensure that all purchased material arrives on time and is 100 percent acceptable. To do so, measurements need to specify exactly what material or service is needed and at what performance level. If the needs are not specified correctly, the performance of the purchased material may not satisfy the requirements to do the job correctly. The measurements for the purchasing process, then, must consider the intent of the purchasing function.

Another aspect of the purchasing process is to monitor suppliers' performance. In other words, do the suppliers have the capability to meet the company's requirements? Suppliers must be qualified first based on their proven capability. Such qualification includes the purchasing department's assessing suppliers' process management systems and evaluating the sample products and financial viability. After a supplier has been initially qualified, its performance needs to be monitored so the purchasing department can provide feedback to ensure continual improvement of performance. Additional goals must be set to improve the performance of the purchasing process in terms of performance of purchased parts or service, total cost of the purchases, response time, and suppliers' service.

A purchasing department might respond to the questions asked above as follows:

1. What is the intent of the process? *To ensure 100 percent acceptability of the purchased products and services when needed.*

2. How do we know the process is producing good output? *We monitor levels and trends in the number of defects or errors found in the operations, total cost of the purchased items, availability of the purchased products or services, and suppliers' service.*

3. Which process input or in-process parameters affect the process output directly? *Process input parameters will be*

the accuracy of requirements supplied to suppliers, the in-process parameters will be the ongoing monitoring of suppliers' process management systems, and the output parameters will be performance of specific parts received.

4. Which process parameters are more critical than others? *The parameters can be prioritized based on the variance in performance and the economic impact of the received material during evaluation or in production stage. Parameters with larger variance and parts with high prices are critical items to monitor.*

5. How do we ensure that the critical parameters are operating at established values? *Capability data must be gathered about parts and performance evaluated with respect to established specifications. If deviation from specification is significant, suppliers' processes must be looked into for potential causes for the deviation.*

6. How will we collect and analyze the data to establish relationships between the process output and the input or in-process data? *A data collection method for selected parts' parameters must be established and statistical analysis performed to understand correlation between process variables and product parameters. In addition, the suppliers' data must be correlated with the production performance to ensure effectiveness of the purchasing process.*

ACTION PLAN FOR PERFORMANCE

From the collection of various departmental measurements, the Six Sigma Business Scorecard measurements are identified. The seven elements of the Six Sigma Business Scorecard are shown in Figure 5-4. For each element, a corporate executive is assigned to establish corresponding measurements. The elements of Leadership and Profitability, Employees and Innovation, and Service and Growth require creativity in establishing measurements. Conventionally, these areas have been ignored in improvement efforts because of their inherent subjectivity. Intent and persistence must prevail, however, and effective measurements must be established.

IMPROVEMENT

INNOVATION

FIGURE 5-4.   Six Sigma Business Scorecard.

The measures become part of the operational review each month and are monitored weekly through the information management system. For each measurement, a clear target and a desired rate of improvement are set over a specified time. The executive team must also establish an effective reporting process to publicize the plan and make it easier to share information with employees. For each category of the Six Sigma Business Scorecard, an initiative or a group of initiatives must be created and planned for using a form or equivalent device, as shown in Figure 5-5.

For example, a business that wants to cut the total spending on sales or the cost of purchasing items by 5 percent in 1 year can establish a purchasing initiative. A team composed of individuals from the purchasing, production, quality, accounting, marketing, and design or development departments convenes to

Corporate Strategic Initiative Planning Form				
Leader: VP_X				
Initiative: Improve Spend-to-Sales Ratio by 10%				
Objective: To reduce cost of goods sold and improve margins.				
Target: Improve suppliers' capability by 30%, and reduce cost of purchases by 10%.				
Strategy: Reengineer the purchasing process to optimize the parts performance and total spend $.				
AI #	Action	Responsibility	Committed Completion Date	Critical Resource Required
Expected Benefits:				
Impact on Profitability:				

FIGURE 5-5.   Corporate strategic initiative planning form.

develop a detailed action plan that will achieve the desired objectives. This plan must be designed to accomplish the objectives and targets well—not merely doing them to cross them off a list.

Thinking about performing a task well as opposed to just doing it leads to a different set of actions and requirements. Critical resources required to complete the task well must be identified. This may include a commitment from engineering to use standard parts instead of unique parts for designs, a commitment from key suppliers to provide leadership in their areas of expertise during product development, or commitment from operations to gather information about the performance of parts and share it with the purchasing department to get accurate measures of suppliers' performance.

CORPORATE PLANS

Once the planning for Six Sigma Business Scorecard measurements is completed in all business process areas, the departmental plans must be compiled into a corporate plan to implement the Six Sigma Business Scorecard. All departmental plans must be reviewed for tradeoff among departments, validation of plans, conflicting resource requirements, and clarity of roles for various individuals in the corporation. The compiled plan will have a summary of measurements supported by departmental plans. The plans must also include goals for dramatic improvement, accountability, and celebration when they are carried out with excellence.

When the Six Sigma Business Scorecard measurements, departmental action plans, and the corporate plan have been developed, the entire organizational structure must be reviewed for alignment with the established objectives. The best people must be assigned tasks to achieve the best results and to mentor the other employees. The compensation and savings sharing plans must be developed. Employee development plans—for everyone from the line workers to management—must also be established.

One of the critical aspects to achieving dramatic results is to challenge all employees to achieve better-than-normal results through exceptional teamwork. Success breeds success, and recognition accelerates success. The leadership must be willing to take extra steps to recognize individuals and team contributions. Publicizing success and giving rewards when excellence is achieved are great motivators for other employees to achieve even better results.

PROGRESS REVIEW

Failure to plan is a plan to fail. The planning process is more important than the plan itself. The plan must be dynamic and amended periodically. While the details of the plan may change, the constancy of purpose must be maintained. The Six

Sigma Business Scorecard plans must be reviewed weekly and monthly, both departmentally and as a business, to ensure everyone is playing a role in contributing toward dramatic improvement and higher profitability. Progress must be publicized to build positive momentum, and failures must be analyzed to prevent recurrence.

Finally, the Business Performance Index must be posted throughout the corporation as a minimal common communication for all employees to monitor their company's performance. Just as employees are accountable to their supervisors, leadership is accountable to employees for corporate performance. In the corporate team, all employees root for the leadership, and the leadership roots for all employees. The elements of the Six Sigma Business Scorecard facilitate such a relationship by requiring mutual accountability. The success of leadership is directly related to the success of employees; therefore, let the employees make the leadership successful through planning, preparation, participation, and perseverance.

TAKE AWAY

1. The executive leadership must create a role for corporate growth, and designate an executive, like Chief Growth Officer, accountable for growth by monitoring external factors and promoting employee innovation.

2. The commitment to sustained profitable growth begins with establishing a business scorecard and a vision for the company.

3. What the vision says is less important than how the vision is communicated, and used to inspire employees.

4. By establishing processes to realize vision and goals, we can create measurement criteria that describe the success factors.

5. Employees' loyalty improves when they accomplish a lot, and that happens when employees are intellectually challenged and recognized for their accomplishments.

6. Performing a business opportunity analysis, to identify areas that adversely affect the profitability and growth, is a good way to ensure measurements have a specific purpose rather than just giving nice-to-know information.

7. From the collection of various departmental measurements, the Business Scorecard measurements are established. Leadership and Profitability, Employees and Innovation, and Service and Growth measure intangibles of the organization.

8. For each measurement, a clear target and a desired rate of improvement are set annually. Critical resources required to complete the task well must be anticipated.

9. One of the critical aspects to achieving dramatic results is to set goals that challenge employees intellectually, and require teamwork.

10. Recognizing exceptional results breeds more exceptional results.

SIX SIGMA BUSINESS SCORECARD DEVELOPMENT

Understanding profitability is key to understanding how the Six Sigma Business Scorecard works. Measurements for profitability have existed in businesses from the start, from bartering or trading to mass production. The basic building block of business is shown in Figure 6-1. When a visionary person starts a new business, that business offers a product or service that is bought or produced and sold, packaged, and delivered for a price. The product or service may then be sold again.

As the spiral of selling and getting paid continues, the business leader envisions producing and delivering more and more. If, with luck and diligent effort, the business model works, the business succeeds and begins to grow. The simple measurements in this model include the cost to buy or build, the cost of selling and delivery, and the revenues earned. Profitability is a by-product of these measurements. Remember, profitability is a function of revenue and cost:

$$\text{Profitability} = f(\text{cost, revenue})$$

where cost and revenues are independent measurements for a business, and profitability is a dependent measurement. In other words, to maximize profitability, a business must maximize the difference between the revenue and cost by focusing

Profit = Revenue − Total Cost

FIGURE 6-1.   Basic business building block.

on both cost and revenue at the same time—not on just one or the other. What a breakthrough! This function is a building block of business management, and business owners must keep it in their subconscious all the time.

As a new business survives or succeeds, it starts growing because of increased demand or the founder's ability to offer additional products or services. The founder can no longer run the business alone and hires the first employee. He or she probably rents some space and buys some tools of the trade. The control of the business now rests in the hands of the employee and the founder or the leader. Although the performance of each person is still clearly visible, the complexity of the business is increasing. The new costs that come with growth are added to the existing costs of the business. These additional costs may be ignored as the focus on the basic equation of profitability continues. However, the basic cost model, and its relationship with profitability, has changed as follows:

$$\text{Profitability} = f[\text{cost(build/buy, building, tools,} \\ \text{employee), revenue}]$$

The profitability is still the difference between the cost and revenue; however, the cost part of the equation has become a little more complicated. As long as the business founder is making money and living comfortably, costs may be assumed to be under control and thus taken for granted. If the business loses customers or demand subsides, business management will then become an issue. On the other hand, if a business is successful, other people try to copy it and create competition.

PROFITABILITY VISIBILITY

When the factor of competition is added, one or more of several things might happen: The market will expand, the price will increase, or profitability will be sacrificed, or the business must be shared by more players. Creative selling strategies, such as financing the sale, marketing, or offering discounts, add even greater complexity to the business model. It suddenly becomes a very complex model of cost and revenue, with many factors that can reduce profitability. The clear mental picture of profitability becomes lost, and the risk of not being profitable increases. The partial profitability equation may look as follows:

$$\text{Profitability} = f[\text{cost (material, suppliers' relationship}$$
$$\text{management, tools, employees' performance, waste,}$$
$$\text{buildings, interest on loans, tools, development, innovation,}$$
$$\text{legal, accounting, equipment, overheads, utilities, . . .),}$$
$$\text{revenue (inside sales, direct sales, distribution, relationship}$$
$$\text{management, marketing, advertising, discounts, . . .)]}$$

Even in this equation that simplifies the factors, unless cost and revenue are highlighted, it is difficult to separate cost from revenue variables. In a real business, small or large, most of these variables occur. Complexity increases even more if a business has multiple locations, multiple leaders, multiple minds (and egos) on the management team, and multiple countries of business. To stay profitable and maximize profitability, a business needs a hierarchical measurement system that highlights the simple relation between profitability, cost, and revenue and makes the components of cost and revenue visible at the point of operation. In other words, there must be an integrated measurement system for various complexity factors contributing to cost and revenue. The cost, revenue, and profitability of the business can then be optimized.

To monitor many of the variables contributing to cost and revenue, businesses establish measurements. As a business expands and hires more employees, new roles are defined and functions are created. Complexity again increases, obscuring the view of profitability; thus new measurements must be added to shed light on the profitability, cost, and revenue relationship. Having a clear line of sight to profitability identifies opportunities for improvement when the profitability targets are not achieved.

MEASUREMENT FAILURES

Most businesses have already implemented so many measurements that employees now complain that they spend more time gathering measurements than doing the work, and little comes of the measurement efforts. Only a few companies have assigned sufficient resources to analyze and use the knowledge gained from measurements. The result is that at many companies, an elaborate measurement system is created; however, no one believes in it, no one supports it, and most important, no one plans to use it well. Ineffective measurement systems can be attributed to many causes, including some of the following:

- Questionable intent or purpose of the measurement system
- Incorrect selection of measurement parameters
- Inadequate or poor data collection
- Ineffective analysis
- Confusing communication of goals and results
- Poorly defined ownership for measurements and the processes
- Lack of accountability, i.e., rewards of performance or consequences of poor performance not established
- Ineffective actions based on the opportunities identified by the analysis of measurements
- An unclear relationship between measurements and business performance
- Lack of commitment to data-based decision making

Here is an example of *confusing* communication of goals and results. Many companies establish their annual performance goals only after the year has already started. They then express their goals for reducing the number of defects on a chart as a flat line: the defect rate they want to achieve that year. The problem is that when managers start measuring actual performance against the stated goal, the goal is not achieved, and the target continues to be missed, in part because of the incorrect depiction of the goal line. By the time employees know what goal they are supposed to meet, they are already used to missing the goal line. Consequently, they develop indifference to the established goal, which then becomes irrelevant. Figure 6-2 shows a typical and preferred depiction of the goal line. By correctly depicting the goal line to illustrate the downward trend in the defect rate that the company is striving for, starting with the current defect rate and ending with the desired rate, employees have clear and achievable goals.

The Six Sigma Business Scorecard attempts to address the above issues and improve the effectiveness of the performance measurement system. Before developing a Six Sigma Business Scorecard, we must understand the purpose and scope of any business scorecard and the measurements.

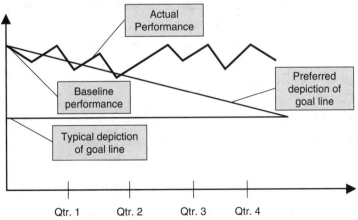

FIGURE 6-2. Understanding significance of defect reduction (goal versus actual).

FEATURES AND BENEFITS OF THE SIX SIGMA BUSINESS SCORECARD

The purpose of the Six Sigma Business Scorecard is twofold: (1) to identify measurements that relate key process measures to a company's profitability, making the opportunities so visible that they are difficult to ignore, and (2) to accelerate the improvement in business performance. Optimizing the profitability, cost, and revenue variables is a primary purpose of the Six Sigma Business Scorecard.

The salient features of the Six Sigma Scorecard include the following:

- It provides a new model for defining a corporate Sigma level.
- It aligns with the organizational structure of the business.
- It maintains visibility of cost, revenue, and profitability.
- It includes leadership accountability and rate of improvement.

The Six Sigma Business Scorecard is a great model for establishing a common target in terms of corporate defects per unit (DPU), defects per million opportunities (DPMO), and Sigma. In the absence of such a target, corporate leaders are unable to focus and suffer from ineffective implementation of Six Sigma. The Business Performance Index (BPIn) provides a one-number measure of corporate performance in terms of Sigma that can be used as a benchmark for driving future improvements. The benefits of the Six Sigma Business Scorecard include the following:

- It provides a target for performance improvement.
- It enables a business to drive dramatic improvement.
- It promotes the intellectual participation of all employees.
- It forces changes in an organization on a continual basis.
- It acts as a catalyst for bringing out the best among employees.
- It generates renewal, energy, and enthusiasm.
- It reduces costs and improves profits.

IMPACT OF CUSTOMER REQUIREMENTS ON MEASUREMENTS

Just like customer expectations, measurements can be subjective and objective. According to the well known Kano's Model, customer expectations have three aspects: assumed, expected, and desired. These customer expectations are shown in Figure 6-3, where the vertical or y axis shows a degree of customer satisfaction and the horizontal or x axis shows the level of effort put into achieving customer satisfaction.

The assumed customer requirements are the basics. For example, someone who goes to the hospital expects to be able to see a doctor there. Someone who buys a bicycle expects that it will come with pedals. If they are there, we ignore them. If they are missing, we complain. In other words, the assumed requirements are communicated only when the customer is dissatisfied.

The expected customer requirements are those that customers have come to anticipate, certain features from their

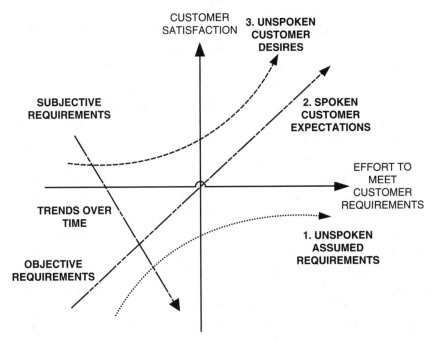

FIGURE 6-3.   Objective and subjective aspects of performance.

experience or by observing them in the marketplace. Customers objectively and clearly state these requirements and pay the supplier for meeting their explicit expectations. The more these requirements are met, the more the customer is satisfied.

The desired customer requirements, however, are not objectively communicated to the supplier. They represent what desires the customer would really like to have met but does not expect. Some call these *customer delights.* Every organization's overriding objective is total customer satisfaction or customer delight. Because consumers always expect more for less (or better, faster, and cheaper), these requirements keep changing. Competitive pressures drive the marketplace requirements for continual improvements in products and services. As a result, features or services that delighted customers in the past eventually become expected. Yesterday's delights become tomorrow's necessities.

There's a method to understanding customers' requirements. Many of us focus on what customers say—not what they mean. We listen more to what we hear than to the unspoken message that underlies the words. Silence can be better than speech, and in the case of customer requirements, the implicit requirements are as critical as the explicit. Why is it important to understand what will delight customers? For a business to be viable, it must fulfill the unspoken or desired requirements. Without this capability, the business will not thrive.

When a business fulfills a customer's spoken requirements, it does not earn any brownie points. Customers see the transaction as a fair exchange of product or service and money. These customers will be very prone to switch to a competing business if there is any slight advantage in doing so. An organization that strives to delight customers, however, has a distinct competitive advantage. This level of customer satisfaction earns customer loyalty and repeat business. These customers even become spokespeople or word-of-mouth advertisers by generating referrals. These customers are partners in the company's success and help lead to innovative solutions with their suppliers. Such relationships lead to better services, innovation, and business growth.

The rewards of delighting customers are so great that a business has great incentive to deliver with excellence the intangible customer requirements, even those implicitly communicated. To ensure that customers are delighted, internal measures must be established, sometimes creatively. While it is difficult to quantify and measure intangibles, effective measurements can be derived with some good thought.

To ensure total customer satisfaction, businesses must measure both the objective and the subjective aspects of performance. While fulfillment of the objective requirements leads to customer satisfaction, the fulfillment of subjective requirements leads to customer loyalty. The more the customers love a company's products or services, the greater their dependability, and the stronger their relationship is to that company. For the customers, this means lower total cost for higher value. The customer is willing to pay premium for additional value, which leads to better performance and higher profit margins.

Objective and subjective measurements can be quantified through attribute or variable data. Attribute data typically have two levels, such as good and bad, pass and fail, low and high, and go and no go. Variable data are preferable because not only do they describe the performance level, but they also quantify it. For example, if service is late for the customer 5 out of 10 times, the next logical question is, By how long is the service late? Instead of recording each order as either on time or late (the attribute data), a company can record the number of hours or days the service is late (the variable datum).

FACTORS INFLUENCING PROFITABILITY

Some functions, such as design, present difficulty in collecting the variable data initially. When greater attention is paid to the details of the design process, however, a business can establish variable data to measure the design process's performance. *The design process is the most influential process in an organization,* as shown in Figure 6-4. *However, it is also the process that is the least measured.* Reasons given to justify the lack of data for the design process include that the function involves creativity and unique projects (no two projects are similar in design), and there just is not enough time to come up with good measures. However, to measure the effectiveness of the design process, a business must take the time to understand the details of the tasks in the design department, the variance or control of the activities, and the significance of each process in the final product. The Six Sigma Business Scorecard addresses the design process as an element of the operational execution category. After all, effective designs of products or services require innovation from all aspects of the business.

To make objective and subjective contributors to profitability more visible, the Six Sigma Business Scorecard ranks and weights categories based on their significance to profitability (see Figure 6-5). According to this model, the leadership has a significant impact on the profitability, which is reflected in the compensation the company's leadership receives. The leadership has direct responsibility for the profitability and should be accountable to employees for its actions and behaviors.

GROWTH AND PROFITABILITY

One of leadership's roles is to inspire employees to contribute to the business intellectually as well as physically through

FIGURE 6-4. Cost influence factors.

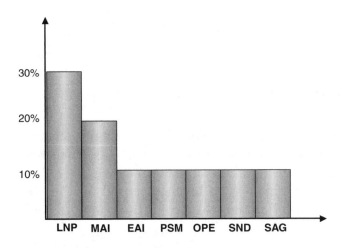

Legend:
LNP => Leadership and Profitability
MAI => Management and Improvement
EAI => Employees and Innovation
PSM => Purchasing and Supplier Management
OPE => Operational Execution
SND => Sales and Distribution
SAG => Service and Growth

FIGURE 6-5. Business Scorecard categories and influence on profitability.

clear objectives and meaningful recognition. Another equally important management responsibility is to ensure that the business processes are improving fast enough. Unless the executive leadership holds this mind-set, corporate processes will be managed inefficiently without improvement. In most businesses, processes are currently managed according to the budget, which is linked to business growth rather than profitability. Departments grow as the budget grows, and with the increased workload, the budget can become unmanageable: The race for growth leads to losing the visibility of the profitability.

No business, irrespective of its size, can sustain growth without profitability. Processes must be challenged frequently so they remain relevant and effective in terms of cost, response time, and performance. Aggressive improvement goals must be set to avoid cost cutting through layoffs or attrition in tough economic times. Setting aggressive improvement goals will maintain the line of sight between business processes and profitability objectives.

OWNERSHIP FOR PERFORMANCE

With the Six Sigma Business Scorecard, a business devises measures for three levels—Leadership, Executives, and Department Managers, as shown in Figure 6-6. These roles at these three levels may vary from business to business based on the size. Small businesses might have the Executives and Process Owners combined, while larger businesses maintain all three levels. The leadership level is the CEO, or the highest position in the company.

The success of the Six Sigma Business Scorecard depends upon how well the leadership owns the performance of the business in the real sense; i.e., their compensation is tied to the performance and profitability of the company. The Business Performance Index (BPIn) provides to leadership a visible measure of business performance. The leadership must know the BPIn on a weekly and monthly basis. Without this knowledge, the profitability for the month cannot be affected and the BPIn cannot be improved.

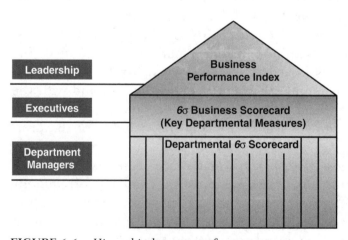

FIGURE 6-6.   Hierarchical structure of measures.

The following shows the frequency of review required to maintain a certain level of profitability:

PROFITABILITY OBJECTIVES	REVIEW CYCLE
3 to 5 years	Annual
Annual	Quarterly
Quarterly	Monthly
Monthly	Weekly
Weekly	Daily

The leadership must also monitor the variance in the business performance—not just performance levels. Reviewing the business performance frequently and communicating that information with the executive team enable the business to take timely remedial actions and achieve the desired business objectives. During the review, the leadership must have very clear expectations for the executive team in terms of the rate of improvement, innovation, and trends for business performance.

The executive team members (or the equivalent in small businesses) are responsible for ensuring optimal performance of the business processes, good interaction between processes, appropriate strategic alignment, good teamwork, and positive trends in performance. Collectively, the executive team is responsible for ensuring that the Six Sigma Business Scorecard is developed and used for monitoring the business performance. Executives review the performance periodically and take appropriate actions to adjust processes as needed to maintain profitability. The information for these measurements is pulled from the process-level measurements.

STEP-BY-STEP DEVELOPMENT OF THE SIX SIGMA BUSINESS SCORECARD

To develop the Six Sigma Business Scorecard, the executive team must outline the business process flow, identifying key processes and establishing measures of their effectiveness. Measurement parameters are selected for critical inputs, in-process, and outputs of the process. The team establishes a total error rate for each process by summing all defects or mistakes and normalizing them to the number of units inspected or verified. The indicators for each process can be summarized in a group of measures including cycle time, process Sigma, and cost or effectiveness.

To calculate Defects Per Million Opportunities (DPMO), the team defines opportunities for errors based on actions that could produce errors. For example, in invoice prepara-tion, the number of items that are manually entered could be counted as the opportunities for error. Departmental Defects per Unit (DPU) or DPMO can be calculated, and Sigma levels can be determined. For nonproduction processes, the formulas for DPU and DPMO are restated in terms of errors instead of defects.

$$DPU = \frac{\text{Number of errors made at a process}}{\text{Total number of units produced in a department}}$$

$$DPMO = \frac{\text{Number of errors made at a process} \times 1,000,000}{\text{Total number of opportunities for making errors}}$$

Following is a list of steps to guide development of the Six Sigma Business Scorecard:

1. Understand the intent of the Six Sigma Business Scorecard.
2. Commit to using the Six Sigma Business Scorecard by integrating Six Sigma in a revised vision of the company.
3. Create a Business Performance Index (BPIn).
4. Establish short-term and long-term improvement goals for a profit center or the company.
5. Establish measurements for each category of the Six Sigma Business Scorecard for each profit center.
6. Establish the relationship between profitability and Six Sigma Business Scorecard measurements.

7. Develop plans to utilize technology to automate the data collection and analysis.

8. If multiple profit centers exist, establish an aggregated Six Sigma Business Scorecard for the corporation.

9. Identify key processes for improving business performance.

10. Identify input, in-process, and output process parameters.

11. Establish data collection methods for these process parameters.

12. Collect data and calculate the error rate, cycle time, and cost for each department.

13. Plot trend charts and present the data with respect to established goals on a weekly basis.

14. Publish the weekly BPIn and monthly Six Sigma Business Scorecard reports.

15. Review business performance using Six Sigma Business Scorecard results.

16. Identify measurements with the greatest variance and adverse performance.

17. Investigate for root causes of the variance and waste in those areas.

18. Initiate remedial actions to improve the performance.

19. Monitor impact on the BPIn and profitability.

SAMPLE MEASUREMENTS

Developing measurements for a process requires a mind-set for accountability. With the spirit of accountability, the team can identify measurements that reflect the performance of the process itself and its relationship with the performance of the company. Measurements must be understandable, quantifiable, and economic. The data must be easy to collect and analyze. The measurements' intended use is to identify opportunities for improvement instead of to punish someone for errors. The process measurements must also be linked to the Six Sigma Business Scorecard. A sample of measurements for various functions is shown in Figure 6-7.

Function	Processes	Measurements
Purchasing	Purchasing Requisition	Number of Errors per Requisition
Sales	Quotation Sales	Total revenue Number of new customers Value of repeat business Number of proposals accepted to total proposals submitted
Engineering	Design	Variance in on-time completion of design Completeness of design outputs Variance of design from target performance Changes after design release Reproducibility of design
Accounting	Accounting	Receivables age Timeliness of financial measures or reports Invoicing errors
Quality	Inspection Audits	Cost of poor quality Number of recurring problems Improvement opportunities
Management	Leadership	Rate of improvement Planning Communication Employees' satisfaction

FIGURE 6-7.   Matrix of measurements.

For each process measurement, the team must establish a trend chart to drive the rate of improvement. Since the goal for rate of improvement is aggressive, the range of values on the *y* axis changes significantly and can become invisible at lower values due to the wide range. The Log Normal trend chart, where each major division represents an order of magnitude instead of a linear increment of the value, can correct this situation. Figure 6-8 shows a chart with the goal line set for 3 years. A similar chart can be made for 1 year as well to magnify presentation of the data.

BPIn EXAMPLE

Based on the process measurements and the Six Sigma Business Scorecard, the customized BPIn is created as shown in Figure 6-9. The BPIn has been standardized for competitive

FIGURE 6-8.   Six Sigma performance tracking chart.

Measurements	Category Abbreviation	Category Significance	Performance Against Plans	Index Contribution
1. Employees' Recognition	LNP	15	60	9
2. Profitability	LNP	15	50	7.5
3. Rate of Improvement	MAI	20	80	16
4. Recommendations per Employee	EAI	10	80	8
5. Total Spend/Sales	PSM	5	60	3
6. Suppliers' Defect Rate	PSM	5	60	3
7. Operational Cycle Time	OPE	5	80	4
8. Operational Sigma	OPE	5	80	4
9. New Business/Total Sales	SND	10	90	9
10. Customer Satisfaction	SAG	10	80	8
Corporate Wellness (BPIn)				71.50%
Corporate DPU				0.3355
Corporate DPMO (number of executives = 9)				37,275
Corporate Sigma				3.27

FIGURE 6-9.   Sample index report card for a medium-size company.

benchmarking. A company can compare its BPIn to its suppliers' BPIn and promote teamwork. An analysis of the BPIn measurements in Figure 6-10 shows that most of the measurements are quantifiable, contain easy-to-collect data, and have a direct relationship to profitability. The two measurements that have an indirect relationship to profitability are employee

BPIn Measurements	Quantifiable	Ease of Data Collection	Relationship to Profitability
1. Employee Recognition	Yes	Need to establish a process	Important for customer satisfaction
2. Profitability	Yes	Yes	NA
3. Rate of Improvement	Yes	Yes	Yes
4. Recommendations per Employee	Yes	Need to establish a process	Critical for business growth
5. Total Spend/Sales	Yes	Yes	Direct relationship to profitability
6. Suppliers' Defect Rate	Yes	Yes	Direct relationship to profitability
7. Operational Cycle Time	Yes	Yes, may take effort	Will impact profitability directly
8. Operational Sigma	Yes	Yes, may be challenging for office functions	Direct relationship to profitability
9. New Business/Total Sales	Yes	Yes	Critical for growth
10. Customer Satisfaction	Yes	Yes	Yes

FIGURE 6-10.   Analysis of BPIn measurements.

recognition and employee innovation. A company with multiple divisions can use the BPIn to develop an independent BPIn for each division and aggregate it for a corporate BPIn.

The Six Sigma Business Scorecard and BPIn can be developed for a business of any size. The BPIn has been standardized to establish a benchmarking process between companies, including suppliers. The Six Sigma Business Scorecard measurements can be modified depending upon the business. The main intent of the Six Sigma Business Scorecard is to drive dramatic improvement internally and improve profitability. Initial application of BPIn at sample companies has resulted in a very acceptable and accurate indication of corporate performance.

TAKE AWAY

1. Cost and revenue in the profitability equation are independent measurements while profitability is a dependent measurement.

2. To maximize profitability, we must maintain cost, revenue, and profitability visible through measurements and actions.

3. Before developing a Business Scorecard, we must understand the purpose and scope of any scorecard and the measurements.

4. The Business Scorecard must make the opportunities so visible that they can not be ignored, and accelerate the improvement in business performance.

5. The Business Scorecard weighs measurements based on their significance of contribution to business performance.

6. The main intent of the Business Scorecard is to drive dramatic improvement internally, and execute the fundamental business strategy of achieving sustained profitable growth.

MEASURES OF INNOVATION

Business scorecards have been practiced for many years. The Balanced Scorecard was introduced to the business world in early 90s. Since then a lot has changed in the business world such as globalization, advent of Internet, outsourcing, and information age. The impact of the changes in last 15 years has led to intensive competition, and demand to innovation faster than ever. As a matter of fact, rate of innovation has been growing for many years; however, the current demand requires sustained acceleration in innovation. Now, there is a need for institutionalizing innovation, and role of innovation is becoming more apparent in the business growth. Thus, there is a demand for measures of innovation.

Academia and corporations have been tinkering with measures of innovation, have developed many indices, but have found insignificant correlation with any existing business performance measure, such as profitability, growth, or market capital. Actually, recent reports in various magazines or studies have shown that investment in innovation shows negative correlation with business performance. How can that be possible? One explanation is that we simply do not understand the innovation process enough to come up with meaningful measures of innovation. However, the work must continue to identify right measures of innovation, and innovation has become an imperative to superior business performance.

In order to have right measures of innovation, we must streamline our measures of business performance. One of the challenges we face today is that most of business measures relate to the financials. Even if the corporations implement balanced scorecard like measurements, the executive thinking is still heavily oriented toward financial measures. This is apparent from the trends on Wall Street, and response in corporate boardrooms to quarterly expectations. In other words, we are trying to manage the quarterly numbers on Wall Street, and trying to fine tune business accordingly, when knowing that quarterly performance on Wall Street has, sometimes, nothing to do with the business performance.

As a first step toward measuring innovation, we must have the right mindset to measurements. Such measurements must proactively identify opportunities for improvement and innovation, and affect profitability and growth. We have seen corporations trying to exceed low expectations of Wall Street by a cent. To increase profits accordingly, employee benefits are adjusted, headcounts are managed, and revenues are manipulated. As long as we will have such an executive mindset, business performance will be a game of dice.

The Six Sigma Business Scorecard represents a first step in the direction of fine tuning business performance for sustaining profitable growth. The scorecard has measures for profit and growth. To maximize profit, cost is emphasized, and to maximize growth, innovation is emphasized. Of course, several measures facilitate both profit and growth. That's how intertwined corporate performance is, which simply cannot be linearly impacted because of nonlinear correlations.

Our understanding of innovation has been limited to brainstorming sessions, and technological evolution. Businesses like 3M have been innovating for decades. However, as a society, due to lack of understanding of the innovation process, the process of innovation has been random, and perceived to be the act of a genius, or a mystery. In order to have meaning measures of innovation, this must change. We, first, must learn about the process of innovation, before we can have useful measures of innovation.

Then, why does one measure innovation? What are the objectives behind measuring innovation? The objectives could be to establish a relationship between innovation and the market capitalization, predictability of the innovation process, or the rate of innovative products; determine performance of the innovation process; and assess availability of necessary resources for the innovation process, or just an overall assessment of the innovation process for an organization. If one considers such questions one by one, we would answer each question differently. Data needed to quantitatively answer the questions would be different, and thus the measurements for innovation would be different.

UNDERSTANDING INNOVATION

Accordingly, to identify measures, one must understand the purpose of innovation, its environment, and the input, in-process, and the output parameters. Further, the relationships between input and output variables of the innovation must be understood. To determine measures of innovation, one must also understand the role of each process in creating the desired innovation. Figure 7-1 shows steps in the innovation cycle which starts from demand for innovation to publicizing innovation. One can see that there can be many measures that can be

| Demand for innovation |
| Generate ideas |
| Develop the innovative solution |
| Commercialize the solution |
| Publicize the innovation |

FIGURE 7-1. Cycle of innovation.

used for monitoring innovation. The challenge is that people want to come up with some magical measures of innovation that can tell the whole story and are predictors of innovation. Most people would like to identify some measures, provide incentives, and start monitoring it. Given the current understanding of the innovation process, one can not establish an adequate and accurate measurement system right away; instead one can start an initial set of measures.

Figure 7-2 shows how one can take a look at the process of innovation. Once again, one can apply the 4P model of process management to understand the innovation. Accordingly, there must be some preparatory steps for innovation, such as having right tools, information, methodology, and people. The tools are required to experiment for developing a new solution, information is needed to gain more knowledge of the subject, methodology facilitates continuity of the innovation process, and people skills are critical in thinking differently and generating ideas. In order to institutionalize innovation in an organization, the leadership must address issues of business strategy for innovation, organization structure, culture and environment for creativity, compensation and incentives, recognition and rewards, and intellectual property protection. For practicing the innovation process, one needs to establish a benchmarking

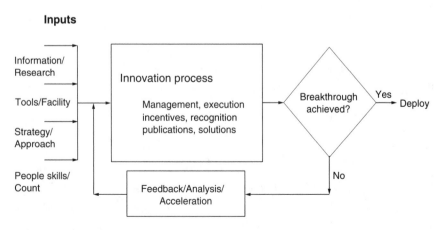

FIGURE 7-2.   Understanding measures of innovation.

process, scout for opportunities for innovation, establish targets for innovation, and provide resources for innovation. Figure 7-3 shows types of measures for measuring effectiveness of the innovation process. It also shows how difficult it can be to come up with measures of innovation due to complexity of the innovation process. To develop a measure of effectiveness of a process, one needs to look into its inputs, activities, and outputs. If one is interested in developing innovation index for an organization, one must consider factors such as differences among organizations, and key processes. Understanding cause and effect relationship between the process output and its inputs helps in identify right measures of effectiveness. In other words, when establishing measures of innovation, one must define purpose and scope, and look into critical inputs, activities and outputs. Based on causative relationship, criticality and controllability of a process step, and practicality of data collection, one can identify some measures to monitor performance and improvement.

Figure 7-3 has classified measures of innovation in two categories: business level and process level. The business level measures are more relevant in terms of the output and the value of the innovation process. The process level measures are more appropriate at the function level. At the input level, business measures include resources for its innovation

Process elements	Business level measures	Process level measures
Input	Funding, culture of risk taking, reward, tools	Excellence in research, innovation management, time allocation (%)
Activities	Targets for innovation, process of innovation, extent of institutionalization, idea management, internal and external publications, knowledge management, internal and external collaboration, recognition	New idea deployment, Extent of improvement or change, Degree of differentiation, disruption or innovativeness, time to innovate
Outputs	Patents, new products, services or solutions; sales growth, market position or ranking, customer perceptions	Rate of innovation, savings, opportunities, potential

FIGURE 7-3.   Types of innovation measures.

process, various major tasks, and expected deliverables to effectively manage the innovation process in an organization. At the process level, measures are limited to a specific task. For example, if the idea management is a critical process for accelerating innovation in an organization, its inputs, activities, and outputs are measured. Inputs to idea management can involve a number of employees or allowed time for innovation related activities. At the activities level, one must consider critical aspects of idea management process, which may include the process of idea collection, analysis, and selection for action. This may also include training employees in the art of generating innovative ideas. This may include effective brain storming sessions, utilizing Internet for gaining knowledge, or collaboration within or outside the company.

MANAGING IDEAS

One of the methods for generating ideas, the author has discovered, includes generating "good," "crazy," "stupid," and "funny" ideas. The good ideas come easily to employees based on what they have seen and done. Most likely, these ideas are tired up one, and are not well thought through. To facilitate innovative thinking, one can extend thinking in sequential manner, i.e., build on the current activities, for generating crazy ideas. Next, one can think in broader terms, sometimes, totally unrelated to the current activities, for stupid ideas; or one can even think of some laughable ideas, which are totally unexpected by extending the sequence of activities and relevance together. It has been found that thinking of funny ideas is the hardest thing to do for most of us, and the funny ideas happen to be the most innovative ideas.

One of the techniques to generate many ideas in a group discussion includes distributing 3 inch × 5 inch cards to everyone, and having participants write their own ideas on the cards with total silence. On a topic, one can ask every one to write three good, crazy, stupid, and funny ideas. By doing so, a team can quickly generate several ideas per participant in a very short time. It has been observed that, in order to generate one practical innovative solution, many ideas need to be generated related to the innovative solution. This implies that an organization must focus on achieving excellence in idea management.

In the past, organizations have tried several versions of the idea management process. For example, many companies had implemented employee "suggestion" or "recommendation" programs. To do this, several idea boxes were mounted with a lock on walls all over, and a small group of volunteers were selected to form a committee to review and manage these ideas. This process had faced challenges due to lack of time committed to this activity due to main job responsibilities of the volunteers. In most cases such suggestion programs failed because the ideas were about getting more bonus, complaints about supervisors, failure of leaders and the list goes on. Employees, instead of using it for improvement purpose, used the suggestion program to communicate with the leadership. The suggestion programs of the past highlighted a problem of communication between management and employees. Most suggestion programs were shut.

In this knowledge age, employee ideas are becoming more critical. We simply can not ignore them. We must listen to all ideas from employees, and provide a vehicle to communicate their inputs for improving business performance. The improvement could be at the activity level, process level, and at the product or service level. Actually, the leadership must do everything in its power to maximize employee participation in the idea generation program. Given most employees participate, the challenge is how to get their innovative ideas without dealing with some social or unrelated issues. It requires establishing guidelines for submitting ideas by highlighting and communicating organizational needs for employee ideas. One can see that it requires more than volunteers' time. The management must be concerned with the return on investment in the idea management.

In order to accelerate innovation, organizations will need many innovative ideas from employees. Ideas lead to activities, activities form a process, processes generate products or services, and products or services generate growth and profit margins. Any investment in management employee ideas can be easily economically justified due to the new opportunities coming out of it. Thus, to establish measures of innovation, one must look into identifying measures for effectiveness of the idea management process. Today, many new software tools are available to manage employee ideas. However, the need for idea management process has not been understood yet, due to lack of understanding of the innovation process, or the right focus in an organization.

DIRECT MEASURES OF INNOVATION

Today, corporations are skeptical of adding new measurements to the existing portfolio of measurements. The current measurements do not, however, get fully used through analysis for extracting business intelligence or creating new opportunities continually. Instead, too many companies end up measuring too many things for too little values. Currently, there are varieties of dashboards and scorecards that focus more on display, instead of extracting intelligence out of the data. These are tools for data mining. The waste continues to pile up without proper understanding of the related processes and a planned application of the lessons learned from the measurements.

Recently conducted research has shown that the correlation between various measures of innovation and its impact on its output has been somewhat limited. Measuring innovation has been challenging because current measures do not provide statistical analysis, or do not tell us with confidence the impact of innovation, or lack of it on wealth creation. Such measures of innovation are not available for strategic planning because of the uncertainty associated with measures of financial impact of innovation. Ultimately, there is a lack of financial, organizational, and cultural structure around innovation.

The Business Performance Index (Gupta, 2003) consists of seven elements and ten measures for monitoring business performance. The three measures out of ten measures are shown in Figure 7-4. Three measurements are CEO recognition for exceptional value creation, employee ideas, and sales growth from innovative products or services. These three measures contribute positively at different stages of the innovation process to institutionalize innovation in an organization. The CEO recognition inspires employees to produce exceptional solutions, tracking employee ideas ensures intellectual engagement of employees, and sales growth ensures economic sense of the innovation process.

It is logically apparent that CEO recognition of employees and incentives act as a catalyst for innovation; idea management is the process; and sales of new products, service, or solution is an output of innovation processes at a corporation. One can easily combine these three or similar measures to develop the corporate index that is simple and measures key aspects. The objective of measures of innovation is to provide trends in performance and identify areas for adjustment to accelerate innovation.

Having understood the innovation process, the measures of innovation, and the elements of innovation indices, one would wonder how to start measuring innovation at their organization. One must not just copy another organization's measures of innovation as they may reflect different priorities to

Six sigma business scorecard elements	Measures of innovation	Innovation process elements
Leadership and profitability	CEO Recognition of employees for exceptional innovation	The critical input to accelerate innovation with a clear purpose and expectation.
Employees and innovation	Employee ideas for improvement and innovation	Idea management is the first process activity towards managing innovation. All innovations begin with an idea.
Sales and distribution	Sales for new products, services, or solutions	Growth in revenue due to innovative products or services is the direct measure of the output of innovation in an organization.

FIGURE 7-4.   Initial measures of innovation.

act and commit resources. Figure 7-3 demonstrates that there can be many measures of innovation. Depending upon the corporate strategy, structure, culture, resources, and objectives, one can identify at least three measures relating to the input, in-process, and output of the innovation process. Anything more than three can require extensive resources, and cause confounding relationship between the value of innovation and resources committed to the innovation process. Good measures of innovation, being specific, measurable, and actionable, catalyze the innovation process and produce significantly more innovative outcomes.

Current corporate measures include indirect measures of innovation. Being a new process evolving toward standardization, difficulties are expected in collecting data for various measures of innovation and benchmarking. However, commitment to institutionalizing and measuring the innovation process must start in order to accelerate innovation in a measurable way. Measuring CEO recognition, number of ideas per employee, and sales for new products and services are three direct measures of innovation to begin with.

TAKE AWAY

1. Organizations need to accelerate innovation for achieving growth.
2. Current measures of innovation are obscure, and do not provide feedback.
3. Innovation process must be understood before establishing meaningful measures of innovation.
4. Too many measures of innovation will lead to minimal correlation between cause and effect of innovation.
5. Three direct measures of innovation are CEO recognition, employee ideas, and sales growth.
6. Combining three can lead to a practical innovation index for an organization.

IMPLEMENTING THE SIX SIGMA BUSINESS SCORECARD

Once corporate leaders decide to adopt the Six Sigma Business Scorecard to monitor corporate performance, they need to understand and communicate the relationship between performance and profitability. The first step in implementing the Six Sigma Business Scorecard is to perform a feasibility study to identify which categories of improvement can be easily implemented and which will require greater effort. A cross-functional team within the organization or an independent group can be assigned to investigate.

Based on the processes that exist in a business, the leadership team must develop a Six Sigma Business Scorecard that is suitable to achieve its process improvement objectives. If any doubts about the effectiveness of the scorecard exist, the Six Sigma Business Scorecard can be validated using available historical or estimated performance data. Initial challenges include devising measurable ways for the leadership to recognize employees for improvement, determining the rate of improvement for each department, quantifying innovation by employees, tracking new business as a portion of total sales, and measuring operational excellence in service or support areas. Existing measurements, such as total expenses, total sales, suppliers' performance, production performance, and customer satisfaction, are easier to quantify. Common current measurements account for about one-half of the performance indicators of BPIn.

The executive team can estimate corporate wellness, determine the Business Performance Index (BPIn), and identify areas to achieve improvement at the highest level. The BPIn and Sigma level can be used as a rough baseline of the corporate performance and as a catalyst to commit to an integrated corporate performance measurement system, the Six Sigma Business Scorecard. The BPIn can be used as a leading indicator of the corporate performance. Using the BPIn creates a strategic intent to implement the Six Sigma Business Scorecard.

CREATING AWARENESS

Having committed to implementing a corporate performance measurement system, both leadership and management must be trained to understand the intent, objectives, scope, and measurements of the Six Sigma Business Scorecard methodology. The methodology includes the following:

- Identifying measurements
- Creating a business growth and profitability model
- Identifying changes in organizational structure
- Establishing departmental or functional goals
- Participating in leadership and employee evaluation
- Integrating the Six Sigma Business Scorecard with the quality or business management system

The objectives are to understand the role of measurements in improving the profitability and growth rates, and to implement an effective corporate performance measurement system.

BUILDING THE BUSINESS MODEL

Each business is different in many ways: in culture, products or services, leadership, customer base, location, business objectives, and so on. Although a standard Six Sigma Business Scorecard that works for every business is impossible, a standard measurement system to benchmark with other businesses for improvement and investors is important. That's why the system is designed so that the BPIn provides a standard set of measures at the highest level, i.e., the CEO level, and the Six Sigma Business Scorecard provides functional measurement guidelines. In a sense, the BPIn includes tier I measurements, and the Six Sigma Business Scorecard includes tier II measurements.

Development of the business model starts with the simplest form of the cost, revenue, and profitability equation:

$$\text{Profitability} = f(\text{cost, revenue})$$

and expands it to the practical expense and income level. This enables the team to identify factors that have an adverse impact on profitability and growth. This model must be clearly understood and easily visualized by all the executives of the business. Executives who take actions based on the business model demonstrate that they are sensitive to concerns about the company's growth and profitability.

ESTABLISHING THE BPIn

When you are creating a business model, starting with the BPIn measurements is the best way to begin because many of these data already exist. In other words, the team can take the historical data, estimate the unknowns in a cross-functional environment, and establish the relationship between the past performance and measurement. If one or two measurements pose a real challenge to the business or if they are irrelevant, those measurements can be substituted for with similar measurements. Executives must buy into the final BPIn measurements, so ultimately the measurements chosen must be workable for the executives. However, the tendency to remove measurements of employee recognition and employee innovation must be avoided.

Since the BPIn score is a sum of all these measurements, a minor variation in one or two measurements does not significantly affect its outcome. Although some of the measurements might be adapted to fit the business, the weighting of various categories must not be changed unless there is a clear factual need to make an adjustment. Changing the ratings could create significant imbalance in the model, harming the correlation between the BPIn and profitability and growth.

Once the team establishes BPIn measurements, the number of executives reporting to the CEO and COO (or the executive staff) must be determined. This information is required to calculate the corporate Sigma level. Using the number of executives to represent the opportunity for errors assigns responsibility for all corporate decisions to the executive team. They own those decisions, and they, as a team, affect the corporate performance. Executive teamwork is critical to corporate success. Total synergy and integration of executive intents, decisions, and actions are required to achieve sustained growth and profitability.

ESTABLISHING SIX SIGMA BUSINESS SCORECARD MEASUREMENTS

Once the BPIn is established, each executive assumes the ownership of one category. The designated executive is responsible for developing a detailed set of measurements for each category and flowing it down to the process level to ensure that each category score is continually improved through actual performance. The focus should be not on the measurements, but on the improvement. The score changes appreciably only if significant improvement is realized. According to the Six Sigma methodology, dramatic improvement is required to achieve a higher Sigma level.

As the measurements are flowed down throughout the organization, teams need to keep in mind that there are three levels: leadership, executive, and process. The executive level measurements are an extension of the BPIn measurements in the seven categories of the Six Sigma Business Scorecard. The total number of measurements depends on the size of an organization. There are about 10 leadership measurements, as included in the BPIn, about 30 executive-level measurements for the various departments, and many more process-level measurements depending upon the organization's size.

In setting up the executive-level measurements, the number of processes in the department must be examined. At the department level, improvement in process performance and cycle time is the main focus. The process performance level must measure how well the process is running and include in-process measurements for reducing variation and waste. Cycle time measurements guide in improving the time needed to run a process cycle. The in-house measurements must correlate to the customer expectations of better, faster, and cheaper. Therefore, the measurements must include quality, cycle time, and waste of resources.

The process measurements can be integrated into the departmental performance measurements, which then continue to be aggregated into the BPIn measurements. Objectives for the process performance measurements are to improve the defect rate by about 90 percent every 2 years and reduce cycle time by about 50 percent every 2 years. For each department, the defect rate reduction goal translates to about 70 percent per year. Each department can have as many measurements as needed; however, some measurements are internal to the department, and others measure the performance outside the department. For example, within a purchasing department, internal measurements may include completeness of the purchase orders, cost of purchasing, and incoming defect rates. External measurements may include suppliers' lead time or service, and quality or availability of parts in production.

In any process, the number of measurements can be high. Given that the main purpose of measurements is to identify opportunities for improvement, the measurements that lead to process improvement are the most important to identify. For example, the sales department can measure any one of the following: number of inquiries, number of quotations, number of sales orders, number of customers, number of repeat orders, number of new orders, sales volume, etc. However, the department should pick those measurements that are relevant to improving the sales process.

To be profitable, a business must be able to get a sales price higher than the cost of the product as well as sales volume to capitalize on the economy of volume. In addition, the number of new sales orders is a good indicator of the strength of the business. Therefore, the ratio of new business to total sales provides a good measurement for the future health of the business. Measuring operational excellence, however, may involve several measurements, including the sales volume and the number of confirmed orders compared with the total orders placed.

Similarly, a multitude of measurements can exist for the leadership: feedback by employees, the ability to inspire creativity, financial performance measurements, and planning effectiveness. The financial measurements alone are numerous and include the growth rate, profitability, cost of goods, return on net assets, return on investments, and debt/equity ratio. The Six Sigma Business Scorecard's BPIn measurements are standardized across the board. However, for internal purposes the measurements that the team selects should relate to management's ability to improve the performance. The seven categories of the Six Sigma Business Scorecard are generic enough to be the same for various companies with different measurements.

One of the challenges in improving business performance is to maintain the effectiveness of improvement efforts. In many companies, thousands of Six Sigma projects have been initiated. However, only about 10 to 20 percent of those projects have been successfully completed. Many of the success stories are attributed to manipulation of the information by consultants or the assigned project leaders. The successful project reflects on the departmental processes and the rate of improvement.

ENSURING DATA COLLECTION CAPABILITY

Sometimes a business avoided taking measurements because doing so required excessive effort and resources. The data collected were unreliable, and no one analyzed and acted upon the data. The economy of measurements must be maintained and balanced against the available resources for data collection and analysis. How many measurements one company has is not important; instead, how many measurements a company effectively utilizes is what counts. When you are implementing the

Six Sigma Business Scorecard, planning for data collection is one of the more important activities. The format, methods, and tools used for data collection must be planned carefully to reduce the cost of data collection.

Good data collection has several attributes. The data must be easily retrievable and usable. Their integrity must be preserved, and damage must be prevented, through proper backups and a data recovery plan. The data analysis methods must include a graphical display of the data in usable format as well as identifying action items and communication methods. Communicating with employees about performance levels and trends is a major aspect of performance improvement. Employees achieve what is expected of them. Clear communication with employees reduces the gap between the "as is" and the "should be" performance levels.

To achieve dramatic improvement in performance, awareness training must be conducted with all employees to establish a common base of understanding of performance expectations. Employees have the potential to produce expected results once they understand the direction from the leadership and their roles in achieving corporate success. The main distractions in achieving corporate success are conflicting priorities and hidden agendas. Corporate leaders must clarify the importance of a corporate strategy and the role of the Six Sigma Business Scorecard in achieving results. Data integrity at all critical processes, timely interpretation, and reaction when necessary are critical to maintain and improve process performance.

Performance levels, trends, and patterns and their significance are important components of data analysis. During the analysis, if one unsatisfactory incidence of performance is noted, it might be attributed to data inaccuracy, human error, or chance error. If two unacceptable data points exist, they represent a potential for change and indicate that the process needs some adjustment. However, if three data points are unacceptable, process change for one or more patterns is required. *Therefore, the rule of thumb for interpreting data is this: One error is a point that can happen randomly, two errors create a line that sets a direction, and three points indicate a pattern that shows a behavior; one is a go (green), two are a caution (yellow), and three are a stop (red)*. In real life we overreact to aberration and then accept the new pattern. In reality, we may be unable to affect chance variation; however, we may be able to change the pattern.

Based on the data analysis, actions must be initiated. In most cases, no one person or department may be able to remedy the unacceptable situation or solve the process problem. A clearly defined goal, a cross-functional team, and a sound problem-solving approach are required.

WAR ON WASTE

H. James Harrington and Kenneth Lomax in *Performance Improvement Methods* (2000) defined a *war on waste* (WOW) approach to improving performance. The authors recommend developing "battle plans" to conduct the war on waste. The battle plans include data analysis weapons, idea generation weapons, decision-making weapons, and action execution weapons. The DMAIC methodology of Six Sigma incorporates some of these tools as well. The challenge here is that too many people know too many tools, but only a few of those people have the guts to practice them. Some of the common roadblocks in using these tools are identified by Harrington and Lomax:

- Lack of time
- Lack of problem ownership
- Lack of problem recognition
- Errors as an acceptable way of life
- Ignorance of the importance of the problem
- Belief that no one can do anything about the problem
- Insufficient allocation of resources by management
- People trying to protect themselves and blaming others
- Belief in "good enough"
- Headhunting managers

To avoid such roadblocks, the leadership must identify what's required for the business to capitalize on improvement opportunities. The steps include making the problem visible, stating an expectation to eliminate the problem, training employees in problem-solving tools, demonstrating an understanding of the problem and modeling the process of correcting it, communicating the solution, establishing a system to verify effective implementation of the solution, and recognizing the contributions of those who helped solve the problem.

Managers must avoid using their experience to quickly arrive at the solution without objectively reviewing the data analysis. Instead, management must establish the expectations and facilitate an innovative solution by employees. Managers can look into the following pitfalls in the problem solution methods using the Six Sigma Business Scorecard:

- Collecting irrelevant data
- Using incorrect data
- Inability to respond in time
- Analysis of partial data
- Highlighting trivial problems
- Analysis paralysis

If one can deter such activities, the problems will be readily solved and benefits will be realized.

MANAGING CHANGE

Any new project starts with a lot of enthusiasm from a few change agents. However, that change must become contagious throughout the organization and contribute to a positive experience for the business as a whole. During any project's life cycle, the starting point is the most pleasant, with its associated fun and festivities. As the project starts, teams are formed, objectives are defined, plans are developed, and progress is made. When a cross-functional team starts working, members from the various departments may have different expectations. They have different priorities and levels of commitment.

The team goes through the famous four stages: *Forming, Storming, Norming,* and *Performing.* During the Forming stage, the members in the team are identified, and their roles are defined toward a common objective. During the Storming stage (initial meetings), the team members present their views and their experiences and assert their wills. This Storming stage might last from one meeting to several meetings.

As everyone on the team is heard, a common understanding of one another's views is developed. With that common understanding, the team's methodology and rules are established, the team leader playing an important role in establishing them. This is the Norming stage. Within the scope of the objectives and span of the team rules, members reach the Performing stage and participate in implementing the desired change, in this case, the Six Sigma Business Scorecard. Once the planned change has been implemented, it is monitored for a period before the team is discontinued.

In the early phase of the change process, team members and the corporation invest resources to develop methods to work together to implement the change. Somewhere after the first quarter of the change process, leadership might become concerned if few results are evident. This unfulfilled expectation leads to dissatisfaction and frustration, among both leaders and team members. A leader with a shallow commitment might intervene at this point and disrupt the change process, changing the team players and redefining the objectives. These actions could be fatal to the strategic initiative and counterproductive to the corporation. At this point, leadership that focuses on short-term goals might even make a change in personnel, firing some individuals, including consultants if they are vulnerable.

This is one reason why leadership embracement of the full concept is important. The leadership must be committed from the outset to the complete cycle of the change process, as shown in Figure 8-1.

As resources are invested in a change process and few tangible results are evident, team morale may slip because their effort appears to be going nowhere. After the halfway point,

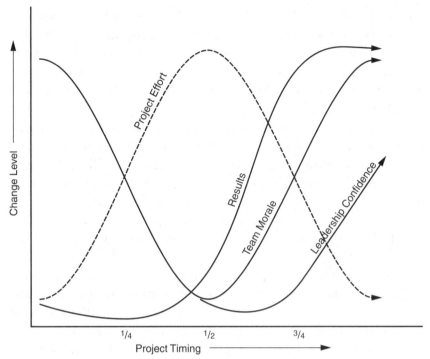

FIGURE 8-1.   The project life cycle.

however, the input effort starts declining and results start showing up. Morale rebuilds and the enthusiasm returns. The leadership must understand that this is a natural change cycle and resist frustration and untimely intervention. Avoiding excessive intervention does not abrogate the leadership from its responsibility to maintain accountability for success of the project.

After the initial success of the pilot projects, the challenge becomes to institutionalize the change. To do so, the leadership must make management responsible for implementing the Six Sigma Business Scorecard and accountable for its success. Each category on the Six Sigma Business Scorecard is assigned to one executive for clear accountability of the initiative and measurements.

To maintain constancy of purpose, focus, and commitment to the Six Sigma Business Scorecard, the leadership must set high expectations for dramatic improvement as well as communicate a sense of urgency. Superior companies maintain the sense of urgency irrespective of corporate performance. Similarly, the strategic initiative of implementing the Six Sigma Business Scorecard must become a priority and be implemented with a sense of urgency to achieve dramatic improvement in corporate performance, including dramatic improvement in profitability and growth.

Initially, some difficulties in implementing the Six Sigma Business Scorecard will exist. However, when its purpose and executive commitment are strong, the scorecard will lead to continual and balanced improvement in corporate performance. At this point, after stable and successful implementation, many businesses tend to either change things because new people have joined the team or create a new initiative to sustain improvement.

Although the Six Sigma Business Scorecard is a dynamic tool that can be revised as the scope and direction of a business change, it is counterproductive to change it simply because of personal preferences or style. When business objectives change, however, the Six Sigma Business Scorecard can be revised to maintain its suitability and effectiveness to the business. Business does not exist to implement the Six Sigma Business Scorecard; instead the Six Sigma Business Scorecard exists to support the business.

INTEGRATING TECHNOLOGY AND THE SIX SIGMA BUSINESS SCORECARD

As the Six Sigma Business Scorecard is being implemented or shortly thereafter, the question often arises about the effort required to maintain it. The primary objective of the Six Sigma Business Scorecard is to facilitate growth and profitability together, ensuring a sound return on investment because of its implementation. However, continual improvement in the Six Sigma Business Scorecard itself must also occur, whether the improvement involves setting new goals, easing the scorecard's maintenance, or integrating measurements with technology.

The value of measurements lies in how they are used rather than in their collection. The value increases with timely response to the information gained from the measurements. The faster a process owner receives feedback about its performance, the better the adjustments made to the process will be. Integrating the Six Sigma Business Scorecard with the corporate information management system can lead to faster response times. Using information technology, the data collection, aggregation, analysis, and reporting of the Six Sigma Business Scorecard can be automated.

Automating data collection and analysis gives the leadership more time to respond to the information instead of wasting time worrying about getting the information. Similarly, process owners can spend their limited resources in improving the process instead of ensuring the data collection for the Six Sigma Business Scorecard. Ultimately, the intent of the Six Sigma Business Scorecard is to drive a *rapid rate* of improvement (not just improvement), experience improved profitability (not just process improvement), and realize business growth (not just revenue monitoring).

TAKE AWAY

1. Perform a feasibility study to identify which measurements of the scorecard can be easily implemented and which will require greater effort.

2. Common current measurements account for about one-half of the performance indicators of BPIn.

3. Starting with the BPIn measurements is the best way to begin improvement in business performance.

4. Given that the main purpose of measurements is to identify opportunities for improvement, the measurements that lead to process improvement are the most important to identify.

5. Automating data collection and analysis gives the leadership more time to respond to the information instead of wasting time and worrying about getting the information.

STRATEGY FOR EXECUTION

Strategy has become an overused word without an action behind it, and results to support it. Kaplan and Norton have stated for over ten years that strategy works less than 10 percent of the time. It sounds like we are making strategy everyday, and executing rarely. It is not far from the truth. Considering reasons for such recurring failures, one must think: Is it the strategy? Is it the implementation? Is it the measure of successful implementation, or simply a case of misnomer? The fundamental business strategy must be "sustained profitable growth" by aligning the organizational vision, with its purpose, customer needs, organization's values, and operational capabilities. The purpose of any strategy is to achieve desired business objectives, which may include customer acquisition, product or process leadership, differentiation, operational excellence, and the intent of being in business.

While strategy is an overused word, very few people talk about execution. Some of the reasons people ignoring execution of strategies are that it requires "real" experience in doing things, it requires understanding of the real-world scenarios, it is a complex issue involving many variables, and, most importantly, it requires accountability. I have not heard many people held accountable for poor strategic plans, versus many operational guys getting axed of for not delivering results. We have seen that strategy guys develop strategies, and operational guys execute strategies, but communication between the strategy and

Note: The author would like to acknowledge Steve DuBrow and Bill Cooper for their contribution to this chapter.

the execution components of an organization is missing. Somehow, the strategy and the execution guys do not click together. Strategy focuses primarily on external issues, and execution focuses primarily on internal issues. In the end, the strategy meets execution, and both fail. The challenge is how to develop strategy for execution. In other words, if an organization develops a strategy that it can execute well, the results will be there. A strategy could stretch operational resources, but the strategy cannot overwhelm the organization's resources to fail its execution.

STRATEGY

In an attempt to understand "strategy" better, review of various definitions on Internet reveals the following elements of a strategy (*http://www.google.com*):

> Strategy is a scheme, a long-term plan of action, a sequence of activities and reactions, a nonspecific statement, a plan for success and achieving an advantage, a recipe to fit separate components together to contribute most advantageously, a systematic plan consciously adapted, a specific way in which something is to be done, a course of action to achieve targets, an approach of the organization for establishing the future environment, and an allocation of the organization's resources to achieve desired business goals.

Thus the strategy is futuristic sequenced statements for achieving desired performance targets utilizing given resources. Strategy operates within the scope of the organization's vision or so called the *fundamental strategy*. Tactics and procedures are used to implement the strategy. Organizations have multiple strategies relating to financing, marketing, operations, human resources, organization itself, measurements, investments, improvement, and rewards positively affecting quality, cost, productivity and morale. All strategies must serve the grand strategy of achieving sustained profitable growth.

The purpose of a strategy is to serve the corporate purpose by matching business objectives and organization resources. Both elements of this equation have uncertainties associated with them. For example, it is difficult to predict how people will end up performing a certain task, and similarly, it is difficult to predict customer's response to a certain product or service. Even more difficult is to predict market acceptance of certain product or service offered by a company. Thus a good strategy must take into account the available resources and drive corporate performance in a predicted direction with given uncertainties. These uncertainties make strategies fail as they have not been anticipated through formulating scenarios.

In many companies, the link between strategy and execution ranges between unclear and nonexistent. While there are many tools, techniques, and consultants who can provide solutions to one or the other, the biggest challenge and the ultimate nirvana is for the two to be clearly integrated. If done properly, this will provide alignment up, down, and across the enterprise.

We have all experienced (some more than others) the impact of the infamous senior executive strategy meetings. These are the ones where executives venture out to the "shabby" 5-star venues and participate in infamous offsite retreats where the intention was to shape the future of their enterprises. While cynically there may have been more focus on golf than strategizing, hopefully more times than not the team returns with a high-level plan for the future. However, the real work begins in integrating this newly established plan with the rest of the organization.

For example, back in the late 90s and early 2000s when the Internet bubble was at its peak, many traditional manufacturing and distribution companies created strategies to sell their products and services over the Internet in support of achieving a mission of continued revenue growth. While being great in concept, the reality was that there was no understanding or appreciation for the fact that their organizations had no core competencies in online selling and distribution. In addition, there were no processes in place within the organization that could own this strategy and to expect their traditional sales and distribution functions to own this was more than unrealistic. This is not to say that many organizations, given time and dedicated resources, could not achieve this capability, but it was not the quick and dirty solution. As they say, "Nothing is that easy."

The old adage, "What came first: the chicken or the egg?" leads us to a similar question, what comes first: strategy or execution? While this may seem like an easy question and an even easier answer, strategy clearly needs to drive execution; many feel that results in the field may not support this. It is widely recognized by academics, authors, and corporate leaders alike that organizations have had excellent, even great, strategic plans but they are rarely executed successfully.

EXECUTION

One of the reasons most strategies remain in the executive offices and do not reach operations is the lack of emphasis on designing executable strategies. Strategies must be evaluated for their execution. The execution depends upon the following factors:

1. Understand the business environment, and available resources
2. Identify explicit relationship between a strategy and financial performance
3. Demonstrate the leadership commitment to making strategies work
4. Pursue buy-in from the middle management to execute the strategy
5. Review recognition and reward ensuring successful execution
6. Understand organizational capabilities and limitations
7. Align the organization and assign accountability for elements of the strategy.
8. Involve employees and the middle management in approving the strategy
9. Establish executive and employee communication
10. Provide resources and tools to ensure successful implementation
11. Measure and monitor progress for execution and results
12. Provide education to employees in enhancing their problems solving skills

One can see that it takes a lot to make strategies work. Looking at the definition of "execution" on the Internet one can find many versions ranging from legal (death) to carrying out a business strategy or a policy. Accordingly, execution of a strategy is an act to carry out a plan in an organization. When executing a strategy, we must develop an action plan that can be carried out. However, challenges remain even after developing an action plan due to internal barriers in an organization. An effective execution of a strategy is a measure of success of the strategy. The execution depends upon the leadership, culture, and capability of an organization. The leadership's focus and expectation, culture of completing tasks well on time, and capability in terms of resources and skills facilitate successful execution of a strategy.

STRATEGY FOR EXECUTION

A Strategy for Execution (SFE) map for a corporation begins with a fundamental strategy, which is sustained profitable growth. The strategy can only be successfully executed for realizing its potential through four interrelated elements as follows:

1. Measurements: The Six Sigma Business Scorecard
2. Profit Improvement: Six Sigma/Lean Methodologies
3. Business Growth: Innovation or Creativity on Demand
4. Sustenance: Process Management (4P)

Without execution there is really no need for strategy. In this case, a strategy or strategic plan becomes nothing more than a fairy tale to be read and entertained by. Execution is the way things get done. It should be noted that every company is executing something.

It is like driving a car or flying plane without any thought of where you are going. You will keep going but you will never get there because there is no destination—nowhere to get to. The reality is that you will run out of fuel at some point. This is similar to people within an organization who will run out of stamina. What happens then is that many times the movement that they are making is for movement's sake only. They are not guided by anything; and this is where the strategy part comes in. Strategy is the guide while execution is the engine.

Continuing with the car analogy, if a company is running on all cylinders and in fact has an interesting/compelling destination, and then it is the ability to execute that will drive their success. Execution comes in the form of leadership, management, processes, understanding, measurement and the ability to want to "roll up your sleeves" and get the "right" things done. Using tools such as the Six Sigma Business Scorecard, you can tell at any point of time how you are performing toward your stated destination.

The strategy for execution begins with a good set of measures ensuring realization of the fundamental strategy. The Six Sigma Business Scorecard measurements identify operational opportunities for improving profit, and achieving growth. Opportunities for profit improvement are addressed using tools like Six Sigma and Lean principles. The increase in revenue growth is achieved through institutionalizing innovation principles. The innovation begins with an idea, and excellence in idea management will ensure continual development of new products or services. The following describes the steps for executing a strategy to achieve sustained profitable growth using the Six Sigma Business Scorecard.

1. Establish a fundamental strategy to achieve Sustained Profitable Growth (SPG)
2. Utilize SFE map to accomplish SPG
3. Perform a Gap assessment using the Six Sigma Business Scorecard, and identify areas for improvement
4. Establish organization's Vision, Beliefs, Goals, and Initiatives (VBGI)
5. Construct a Business Processes Flowchart, and identify critical processes and their measures of effectiveness. For example, in an engineering services business, proposal

generation is a critical process because accuracy of estimates determines margins at the completion of the project

6. Map the Business Scorecard elements to business processes

7. Align the organization, and establish responsibility to accomplish the vision, and implement the Six Sigma Business Scorecard

8. Measure performance, summarize data, display performance, and communicate expectations continually to all employees

9. Identify elements of the Business Scorecard, and areas in the organization for necessary actions impacting profit and growth positively

10. To improve profit, utilize tools like Six Sigma, DMAIC (Define, Measure, Analyze, Improve, and Control) methodology, or Lean principles

11. To increase revenue growth, emphasize creativity and innovation, which begins with excellence in idea management. Find ways to intellectually engage employees for developing new processes, products, or services

12. Sustain improvement in business performance through the 4P model of process management for striving toward target performance at critical processes

ORGANIZATIONAL ALIGNMENT

There is a belief that people cannot scale but teams can. An individual's ability to execute is finite, while a team's is relatively infinite. The value of a team to get things done and execute is far more effective than an individual's. Simply, a team brings different points of view, skill sets, and unlimited energy to achieve any specific goal.

However, the challenge with the team (and this can be extrapolated to an entire organization) is that, unless they are in alignment, their ability to achieve the goal is at risk. Let us go back to a transportation example for the moment. There is a team consisting of ten members. The team's charter is to navigate from Boston, Massachusetts, in the United States to Sydney, Australia. How will they execute? What obstacles are in their way? What resources do they have available to them? How will they execute? It would not be far fetched to say that different people on the team will have different ideas of how to get the job done.

If the team cannot get "true" alignment on how to get there, some will take different forms of transportation (boat versus plane), some will even go in a total different direction to get there, and some will go west while others will go east. Now without any measures like how long should it take to guide the team, the team may not fulfill their charter in an effective manner. Therefore the team needs to get aligned on the best way to achieve the goal. The true alignment is important because what some team members agree to during planning, is not always the way they execute. This may have to do with culture and differences in understanding, but it is more about people saying one thing and doing something else, based primarily on their own personal self-interest. "I did it this way because it was better for my division and it also allowed me to achieve my personal bonus." At the end of the day, if the team is not aligned on what to do and how to get it done, then the effort it will take to get the job done (if they get it done at all) will be much greater than that with the true alignment. One can envision nine members of a 10-person team waiting in Sydney for the last team member to show up, and they have no idea where this person is or how this person might be coming. The reality is, until the last member of the team shows up, the job is not complete.

This example is easily transferred to any organization attempting to achieve its strategic objectives. If the CFO thinks that to achieve a particular strategy he or she needs to move his or her team in one direction while the VP of manufacturing moves his team in another way, without coordination and alignment, then the odds of achievement become statistically very low. That is even if they clearly see and understand the objective in the same light. Therefore, it is critical to communicate not only during inception of the objective and the initial planning to achieve (the

FIGURE 9-1. A software output displaying elements of the Business Scorecard.

famous offsite) but continually throughout execution. This is again where measures and the Business Scorecard can be used in a fashion similar to a speedometer on a car that communicates how the driver is doing in achieving or getting to the destination. The dashboard displaying critical information gas, speed, distance, and temperature allows the driver to evaluate a situation, and make necessary adjustments in real time. Figure 9-1 shows how a software can provide similar dashboard capabilities to monitor the organizational performance.

BUSINESS SCORECARD FOR STRATEGY EXECUTION

While organizations have historically utilized various methods to measure their strategies, a recent study conducted by Bernard Marr of the Cranfield University School of Management stated it best by saying,

"If organizations do not extract new insights that allow them to make better decisions, then the entire process of Business Performance Management(BPM) is a non–value-added activity or put bluntly: a waste of time and money."

Mr. Marr's point is right on why go to the effort and measure our strategic performance if we are not going to use the information to execute more effectively. He also concludes in his study that companies who formally practice BPM report better execution. If Mr. Marr's findings are accurate, then why are not more organizations practicing formal BPM? Is it too hard? General belief is that BPM is not too hard (though there are challenging components), but rather it is difficult to get started. This is where utilizing the Business Scorecard is most helpful. We speak to hundreds of executives every year on how they are doing in "linking their strategies with execution." After the obligatory response of "we are doing fine," it is clear upon further probing that they do not know what to measure, and where to start. The Six Sigma Business Scorecard can be used to get started. It begins the organizational learning process. Scorecards continue to change as new opportunities to improve are identified. Many times, the final version might be vastly different from the initial Business Scorecard.

The Six Sigma Business Scorecard acts as a catalyst and gets executives thinking about how to measure success within their organizations. One of the challenges in getting started is prior success in managing business performance ad hoc. Their attitude (maybe rightfully so) is: why should I do things differently if I have been "this" successful? Executives must realize that what has worked in the past is not going to work in the future. Business models are changing, global competition is growing, and margins are shrinking. In order to continually make profits, and invest in growth, executives must fine tune the business performance. The gross margin has become the final margin. Without such an understanding, executives can easily lead an organization to extinction. It has only been over the last ten years where a small business in Russia could be selling products to consumers in Boise and Idaho, all with a "click of the mouse." And by the way, this product may not even be manufactured, assembled and/or physically distributed by this entity. It is the power of access to information, and access to and relationship with current and potential customers. The proverbial bottom line is that suppliers and customers have unlimited options, and leaders need to monitor and understand this paradigm to compete successfully today and in the future. In conclusion, the Business Scorecard can be the catalyst for successful execution of the fundamental strategy.

IMPROVING PERFORMANCE

So we leverage the outputs of the Business Scorecard, and find out much to our dismay that we are not executing as effectively as we would like or should within the management and improvement, and/or the sales and distribution areas. The scorecard tells us this but that is just the beginning. We need to do something (execute) to achieve success. Strategies do not get achieved by themselves.

The key is, first, to understand what needs to be done and then to utilize or initiate the appropriate tool/process or methodology to make it better. This could be utilizing the (DMAIC) methodology from Six Sigma or something even simpler.

Second, and maybe more importantly, the organization has to appreciate the effort required to actually execute a strategy. This includes funds, people, training, and time. The biggest failure we experience is where an organization does not dedicate their best people to drive their strategic initiatives. Leaders intellectually want execution to occur but do not in-fact lead the way by dedicating the appropriate resources by giving people enough time and freedom to utilize their creativity to solve the various challenges. It makes us wonder how one can ensure significance of a strategy to the leaders. We already said strategies are not realized by less qualified employees in their spare time. Expecting excessive and inefficiently deployed employee hours, taking time away from their families, is not going to make the strategy work. Even, planning for such resources to achieve success for the organization is downright ludicrous.

THE EXECUTION MAP

In 2001, Kaplan and Norton introduced strategy maps to establish cause and effect logic connecting strategic outcomes with the drivers. In their book, *The Strategy-Focused Organization*, the authors define strategy map as, "A balanced scorecard strategy map is a generic architecture for describing a strategy." The fundamental objective for designing various strategies is to increase the shareholder value. Here, the architecture of the Balanced Scorecard strategy maps consists of the strategy, four perspectives (financial, customer, internal, and learning), measurements, processes, and resources. Strategy maps leave gaps in the value stream for creating value for the customer. Strategy to the four perspectives and process relationship is driven by the numbers instead of the business processes, thus creating disconnects in execution.

According to Dr. Kenneth R. Thompson, Professor of Management at Kellstadt Graduate School of Business, DePaul University,

> "We had some challenges with our clients in their use of the traditional scorecard and in their use of the strategic mapping process (Norton and Kaplan, HBR 2000). They tried to adapt their strategic plan to the scorecard's four perspectives and align them as was advocated in the Norton and Kaplan (2000) article. It was not a good fit for the service-based government organization that we were providing assistance. There was poor alignment and confusion among staffs. We had to convince the organization that the scorecard should be modified to meet the needs of the organization for an aligned set of strategies and measures, rather than trying to modify the organization goals to meet the form of the balanced scorecard."

To improve execution of a strategy, the Six Sigma Business Scorecard is implemented using the execution map, which is defined as a tool to drive business performance through process-based measurements, opportunities, action, and responsibilities. The streamlined execution maps are easy to understand, and do not require extensive outside support for effectively implementing a business scorecard. The execution map consists of the following items addressed through a set of forms as shown at the end of this chapter:

1. Strategy for Execution (SFE) Map

2. Scorecard Execution Roadmap

3. Business Scorecard—Present State

Measurements*	Score (%) (A)	Significance (B)	Net Score = A x B/ 100
1. Employees Recognition		15	
2. Profitability		15	
3. Rate of Improvement in Process Performance		20	
4. Recommendations per Employee		10	
5. Total spend/Sales		5	
6. Suppliers' Defect Rate		5	
7. Operational Cycle Time Variance		5	
8. Operational Sigma		5	
9. New Businesses/ Total Sales		10	
10. Customer Satisfaction		10	
Corporate Wellness (BPIn)		Sum =	
Corporate DPU (use BPIn to DPU table)			
Corporate DPMO = (Corporate DPU/ # of Executives reporting to CEO/COO) x 1000,000			
Corporate Sigma (Refer to Corporate DPMO to Sigma Table)			

Use next several slides to indirectly determine present state for BPIn measurements

4. Ten BPIn Measurements—Indirect Measurements

A. Employee Recognition–Present State

Indirect Measurements	Initial Performance Guidelines	Score (%) (A)	Significance (B)	Net Score = A x B/ 100
% Employees Rewarded for Exceptional Performance	<0.2%–25 0.5%–50 2%–75 >5%–100		45	
Best Company to Work for Ranking by Professional Bodies such as Forbes, Fortune.	Last Quartile - 30 Third Quartile - 50 Second Quartile - 80 First Quartile - 100		15	
Employees' Voluntary Turnover	>10%–50 5%–60 2%–80 <1%–100		20	
Average Annual Applicants as a percent of Employees	<1–50 2–60 5–70 >10–100		10	
Average Training hours per Employee	<1–30 8–50 16–75 >24–100		10	
Score:				

B. Profitability–Present State

Indirect Measurements	Initial Performance Guidelines	Score (%) (A)	Significance (B)	Net Score = A x B/ 100
Net Income as percent of Sales	<0%–0 1%–25 2%–50 4%–60 8%–80 >12%–100		60	
EBITDA	<0%–0 2%–10 5%–30 10%–50 15%–60 20%–80 >25%–100		20	
EPS	<0.5–50 1.0–65 1.5–85 >2.0–100		20	
Score:				

C. Rate of Improvement–Present State

Indirect Measurements	Initial Performance Guidelines	Score (%) (A)	Significance (B)	Net Score = A x B/ 100
Change in SG&A as percent of Sales	<0%–0 0.5%–30 2%–75 >5%–100		30	
Change in CGS as percent of Sales	<0%–0 0.5%–30 2%–75 >5%–100		30	
Change in EPS	<0%–0 0.5%–30 2%–75 >5%–100		30	
Number of Improvements Initiated	<10–15 30–35 50 – 55 80–85 >100–100		10	
Score:				

D. Employee Recommendations–Present State

Indirect Measurements	Initial Performance Guidelines	Score (%) (A)	Significance (B)	Net Score = A x B/ 100
Suggestions per Employee as per the Suggestion Scheme	<0.5–50 1–60 2–70 >5–100		40	
Employee Suggestions in the Business Meetings	<5–10 30–50 50–75 >60–100		35	
Percent of Improvements Initiated by the Employees.	<10%–25 30%–50 50%–75 >75%–100		25	
Score:				

E. Total Spend/Sale–Present State

Indirect Measurements	Initial Performance Guidelines	Score (%) (A)	Significance (B)	Net Score = A x B/ 100
SG & A as percent of Sales	>30%–20 25%–30 15%–60 <10%–100		50	
CGS as percent of Sales	>60%–30 45%–50 35%–75 <25%–100		30	
Ratio of Net Income and EBITDA	<50%–50 60%–60 80%–80 >90%–100		20	
Score:				

F. Supplier's Defect Rate–Present State

Indirect Measurements	Initial Performance Guidelines	Score (%) (A)	Significance (B)	Net Score = A x B/ 100
Percent Shipments Rejected	>5%–0 2%–20 0.5%–50 0.25%–75 <0.1%–100		35	
Percent Deliveries Received On-Time	<75% - 0 80% - 40 85% - 60 90% - 80 >95%–100		25	
Number of Corrective Actions Initiated by the Supplier as per the Organizational Initiative	<5–10 10–40 15–60 >20–100		40	
Score:				

G. Operational Cycle Time Variance–Present State

Measurements	Initial Performance Guidelines	Score (%) (A)	Significance (B)	Net Score = A x B/ 100
Operational Cycle Time Variance for Core Processes	>50% - 25 40%–50 25%–75 <10%–100		70	
Operational Cycle Time Variance for Noncore Processes	>50% - 25 40%–50 25%–75 <10%–100		30	
Score:				

List the Core Processes:

1. _____

2. _____

3. _____

4. _____

5. _____

6. _____

7. _____

List the Key Noncore Processes:

1. _____

2. _____

3. _____

4. _____

5. _____

6. _____

7. _____

H. Operational Sigma–Present State

Measurements	Initial Performance Guidelines	Score (%) (A)	Significance (B)	Net Score = A x B/ 100
Sigma Value for Core Processes	$<3\sigma$–25 4σ–50 5σ–75 $>6\sigma$–100		70	
Sigma Value for Noncore Processes	$<3\sigma$–25 4σ–50 5σ–75 $>6\sigma$–100		30	
Score:				

List the Core Processes:

1. _____

2. _____

3. _____

4. _____

5. _____

6. _____

7. _____

List the Key Noncore Processes:

1. _____

2. _____

3. _____

4. _____

5. _____

6. _____

7. _____

Operational Sigma Calculation Sheet for Core Processes

Core Processes	# of Output Units (A)	# of Defects (B)	DPU (C=B/A)	# of Opps. (D)	DPMO = C x 1,000,000 / D	Sigma Value (From Table)
Total						

Operational Sigma Calculation Sheet for Noncore Processes

Core Processes	# of Output Units (A)	# of Defects (B)	DPU (C=B/A)	# of Opps. (D)	DPMO = C x 1,000,000 / D	Sigma Value (From Table)
Total						

I. New Business/Sales–Present State

Indirect Measurements	Initial Performance Guidelines	Score (%) (A)	Significance (B)	Net Score = A x B/ 100
New Business/Sales	<20%–25 30%–50 40%–75 >50%–100		75	
Potential Sale of New Products (Under Development) as percent of Sale	<20%–25 30%–50 40%–75 >50%–100		25	
Score:				

List New Businesses:

1. _____
2. _____
3. _____
4. _____
5. _____
6. _____
7. _____
8. _____

List Potential New Businesses:

1. _____
2. _____
3. _____
4. _____
5. _____
6. _____
7. _____
8. _____

J. Customer Satisfaction–Present State

Indirect Measurements	Initial Performance Guidelines	Score (%) (A)	Significance (B)	Net Score = A x B/ 100
Customer On - Time Delivery	<80% - 60 85% - 70 90% - 80 100% - 90		20	
Customer Satisfaction with Quality	<80% - 60 85% - 70 90% - 80 100% - 90		20	
Customer Satisfaction with Price	<50% - 60 70% - 100 90% - 70 100% - 60		20	
Customer Initiated Improvement Implementation and Communication	<80% - 60 85% - 70 90% - 80 100% - 90		15	
Customer Satisfaction with the New Products	<80% - 60 85% - 70 90% - 80 100% - 90		15	
Customer Communication regarding Defects with Customer Supplied Products	<80% - 60 85% - 70 90% - 80 100% - 90		10	
SCORE:				

5. Business Scorecard Barometer—Present State

1. Employee Recognition
2. Profitability
3. Rate of Improvement
4. Recommendation/Employee
5. Total Spend/Sale
6. Supplier's Defect Rate
7. Operational Cycle Time Variance
8. Operational Sigma
9. New Business/Total Sales
10. Customer Satisfaction

6. Business Scorecard—Gap Analysis

Measurements	Present Score	Desired Score	Gap/ Comments
Employees Recognition			
Profitability			
Rate of Improvement in Process Performance			
Recommendations per Employee			
Total spend/Sales			
Suppliers' Defect Rate			
Operational Cycle Time Variance			
Operational Sigma			
New Businesses/Total Sales			
Customer Satisfaction			

7. BPIn to DPU Conversion Table

BPIn	0.000	0.002	0.004	0.006	0.008	0.010	0.012	0.014	0.016	0.018
0.00	NA	6.2146	5.5215	5.1160	4.8283	4.6052	4.4228	4.2687	4.1352	4.0174
0.02	3.9120	3.8167	3.7297	3.6497	3.5756	3.5066	3.4420	3.3814	3.3242	3.2702
0.04	3.2189	3.1701	3.1236	3.0791	3.0366	2.9957	2.9565	2.9188	2.8824	2.8473
0.06	2.8134	2.7806	2.7489	2.7181	2.6882	2.6593	2.6311	2.6037	2.5770	2.5510
0.08	2.5257	2.5010	2.4769	2.4534	2.4304	2.4079	2.3860	2.3645	2.3434	2.3228
0.10	2.3026	2.2828	2.2634	2.2443	2.2256	2.2073	2.1893	2.1716	2.1542	2.1371
0.12	2.1203	2.1037	2.0875	2.0715	2.0557	2.0402	2.0250	2.0099	1.9951	1.9805
0.14	1.9661	1.9519	1.9379	1.9241	1.9105	1.8971	1.8839	1.8708	1.8579	1.8452
0.16	1.8326	1.8202	1.8079	1.7958	1.7838	1.7720	1.7603	1.7487	1.7373	1.7260
0.18	1.7148	1.7037	1.6928	1.6820	1.6713	1.6607	1.6503	1.6399	1.6296	1.6195
0.20	1.6094	1.5995	1.5896	1.5799	1.5702	1.5606	1.5512	1.5418	1.5325	1.5233
0.22	1.5141	1.5051	1.4961	1.4872	1.4784	1.4697	1.4610	1.4524	1.4439	1.4355
0.24	1.4271	1.4188	1.4106	1.4024	1.3943	1.3863	1.3783	1.3704	1.3626	1.3548
0.26	1.3471	1.3394	1.3318	1.3243	1.3168	1.3093	1.3020	1.2946	1.2874	1.2801
0.28	1.2730	1.2658	1.2588	1.2518	1.2448	1.2379	1.2310	1.2242	1.2174	1.2107
0.30	1.2040	1.1973	1.1907	1.1842	1.1777	1.1712	1.1648	1.1584	1.1520	1.1457
0.32	1.1394	1.1332	1.1270	1.1209	1.1147	1.1087	1.1026	1.0966	1.0906	1.0847
0.34	1.0788	1.0729	1.0671	1.0613	1.0556	1.0498	1.0441	1.0385	1.0328	1.0272
0.36	1.0217	1.0161	1.0106	1.0051	0.9997	0.9943	0.9889	0.9835	0.9782	0.9729
0.38	0.9676	0.9623	0.9571	0.9519	0.9467	0.9416	0.9365	0.9314	0.9263	0.9213
0.40	0.9163	0.9113	0.9063	0.9014	0.8965	0.8916	0.8867	0.8819	0.8771	0.8723
0.42	0.8675	0.8627	0.8580	0.8533	0.8486	0.8440	0.8393	0.8347	0.8301	0.8255
0.44	0.8210	0.8164	0.8119	0.8074	0.8030	0.7985	0.7941	0.7897	0.7853	0.7809
0.46	0.7765	0.7722	0.7679	0.7636	0.7593	0.7550	0.7508	0.7465	0.7423	0.7381
0.48	0.7340	0.7298	0.7257	0.7215	0.7174	0.7133	0.7093	0.7052	0.7012	0.6972
0.50	0.6931	0.6892	0.6852	0.6812	0.6773	0.6733	0.6694	0.6655	0.6616	0.6578
0.52	0.6539	0.6501	0.6463	0.6425	0.6387	0.6349	0.6311	0.6274	0.6236	0.6199
0.54	0.6162	0.6125	0.6088	0.6051	0.6015	0.5978	0.5942	0.5906	0.5870	0.5834
0.56	0.5798	0.5763	0.5727	0.5692	0.5656	0.5621	0.5586	0.5551	0.5516	0.5482
0.58	0.5447	0.5413	0.5379	0.5344	0.5310	0.5276	0.5242	0.5209	0.5175	0.5142
0.60	0.5108	0.5075	0.5042	0.5009	0.4976	0.4943	0.4910	0.4878	0.4845	0.4813
0.62	0.4780	0.4748	0.4716	0.4684	0.4652	0.4620	0.4589	0.4557	0.4526	0.4494
0.64	0.4463	0.4432	0.4401	0.4370	0.4339	0.4308	0.4277	0.4246	0.4216	0.4186
0.66	0.4155	0.4125	0.4095	0.4065	0.4035	0.4005	0.3975	0.3945	0.3916	0.3886
0.68	0.3857	0.3827	0.3798	0.3769	0.3740	0.3711	0.3682	0.3653	0.3624	0.3595
0.70	0.3567	0.3538	0.3510	0.3481	0.3453	0.3425	0.3397	0.3369	0.3341	0.3313
0.72	0.3285	0.3257	0.3230	0.3202	0.3175	0.3147	0.3120	0.3092	0.3065	0.3038
0.74	0.3011	0.2984	0.2957	0.2930	0.2904	0.2877	0.2850	0.2824	0.2797	0.2771
0.76	0.2744	0.2718	0.2692	0.2666	0.2640	0.2614	0.2588	0.2562	0.2536	0.2510
0.78	0.2485	0.2459	0.2433	0.2408	0.2383	0.2357	0.2332	0.2307	0.2282	0.2256
0.80	0.2231	0.2206	0.2182	0.2157	0.2132	0.2107	0.2083	0.2058	0.2033	0.2009
0.82	0.1985	0.1960	0.1936	0.1912	0.1887	0.1863	0.1839	0.1815	0.1791	0.1767
0.84	0.1744	0.1720	0.1696	0.1672	0.1649	0.1625	0.1602	0.1578	0.1555	0.1532
0.86	0.1508	0.1485	0.1462	0.1439	0.1416	0.1393	0.1370	0.1347	0.1324	0.1301
0.88	0.1278	0.1256	0.1233	0.1210	0.1188	0.1165	0.1143	0.1120	0.1098	0.1076
0.90	0.1054	0.1031	0.1009	0.0987	0.0965	0.0943	0.0921	0.0899	0.0877	0.0856
0.92	0.0834	0.0812	0.0790	0.0769	0.0747	0.0726	0.0704	0.0683	0.0661	0.0640
0.94	0.0619	0.0598	0.0576	0.0555	0.0534	0.0513	0.0492	0.0471	0.0450	0.0429
0.96	0.0408	0.0387	0.0367	0.0346	0.0325	0.0305	0.0284	0.0263	0.0243	0.0222
0.98	0.0202	0.0182	0.0161	0.0141	0.0121	0.0101	0.0080	0.0060	0.0040	0.0020

8. Corporate DPMO and Sigma Table

Corporate DPMO	Sigma	Corporate DPMO	Sigma	Corporate DPMO	Sigma
66807	3.0	6210	4.0	233	5.0
54799	3.1	4661	4.1	159	5.1
44565	3.2	3467	4.2	108	5.2
35930	3.3	2555	4.3	72	5.3
28710	3.4	1866	4.4	48	5.4
22750	3.5	1350	4.5	32	5.5
17865	3.6	968	4.6	21	5.6
13904	3.7	687	4.7	13	5.7
10724	3.8	483	4.8	9	5.8
8198	3.9	337	4.9	5	5.9
				3.4	6.0

9. Vision, Beliefs, Goals, and Initiatives

Vision (One statement)	1-

Beliefs/ Values (Up to 3 phrases)	1-
	2-
	3-

Goals (Up to 5 items)	1-
	2-
	3-
	4-
	5-

Initiatives (Up to 7 tactics)	1-
	2-
	3-
	4-
	5-
	6-
	7-

10. Business Processes

Major Business Processes	Customer	Supplier	Measures of Effectiveness	Process Owners

11. Business Scorecard to Business Process Map

Business Scorecard Element	Key Processes	Process Owners	Supporting Processes	Process Owners
Leadership & Profitability				
Sales and Distribution				
Service and Growth				
Management and Improvement				
Employee and Innovation				
Operational Execution				
Purchasing and Supplier Management				

12. Ten Action Plans to Improve BPIn Measurements

A. Action Plan to Improve Employee Recognition

Leader:

Initiative:

Objective:

Target:

Strategy:

AI #	Action	Responsibility	Committed Completion Date	Critical Resource Required

B. Action Plan to Improve Profitability

Leader:

Initiative:

Objective:

Target:

Strategy:

AI #	Action	Responsibility	Committed Completion Date	Critical Resource Required

C. Action Plan to Improve Rate of Improvement

Leader:
Initiative:
Objective:
Target:
Strategy:

AI # Action	Responsibility	Committed Completion Date	Critical Resource Required

D. Action Plan to Improve Employee Recommendation

Leader:
Initiative:
Objective:
Target:
Strategy:

AI # Action	Responsibility	Committed Completion Date	Critical Resource Required

E. Action Plan to Improve Total Spend/Sales

Leader:

Initiative:

Objective:

Target:

Strategy:

AI # Action	Responsibility	Committed Completion Date	Critical Resource Required

F. Action Plan to Improve Supplier's Defect Rate

Leader:

Initiative:

Objective:

Target:

Strategy:

AI # Action	Responsibility	Committed Completion Date	Critical Resource Required

G. Action Plan to Improve Operational Cycle Time

Leader:
Initiative:
Objective:
Target:
Strategy:

AI # Action	Responsibility	Committed Completion Date	Critical Resource Required

H. Action Plan to Improve Operational Sigma

Leader:
Initiative:
Objective:
Target:
Strategy:

AI # Action	Responsibility	Committed Completion Date	Critical Resource Required

I. Action Plan to Improve New Business/Total Sales

Leader:
Initiative:
Objective:
Target:
Strategy:

AI #	Action	Responsibility	Committed Completion Date	Critical Resource Required

J. Action Plan to Improve Customer Satisfaction

Leader:
Initiative:
Objective:
Target:
Strategy:

AI #	Action	Responsibility	Committed Completion Date	Critical Resource Required

13. Monthly Business Scorecard

Jan, Feb, Mar, Apr, May, Jun, Jul, Aug., Sep., Oct., Nov., Dec.

Measurements*	Score (%) (A)	Significance (B)	Net Score = A x B/100
1. Employees Recognition		15	
2. Profitability		15	
3. Rate of Improvement in Process Performance		20	
4. Recommendations per Employee		10	
5. Total Spend / Sales		5	
6. Suppliers' Defect Rate		5	
7. Operational Cycle Time Variance		5	
8. Operational Sigma		5	
9. New Businesses/ Total Sales		10	
10. Customer Satisfaction		10	
Corporate Wellness (BPIn)		Sum =	
Corporate DPU (Use BPIn to DPU Table)			
Corporate DPMO = (Corporate DPU/ # of Executives Reporting to CEO/COO) x 1,000,000			
Corporate Sigma (Refer to Corporate DPMO to Sigma Table)			

14. Monthly BPIn Tracker

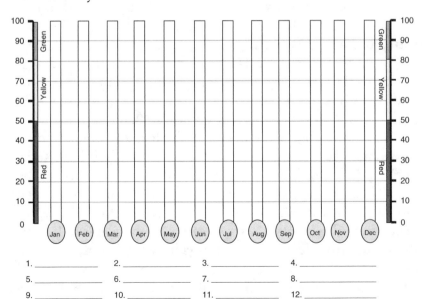

1. _____ 2. _____ 3. _____ 4. _____

5. _____ 6. _____ 7. _____ 8. _____

9. _____ 10. _____ 11. _____ 12. _____

The SFE map identifies the four critical aspects to drive business performance. The four aspects are measurements, improvement, innovation, and management. Any business scorecard cannot produce results by itself. Instead, a scorecard can only identify opportunities for changes or adjustments to be made for maximizing business performance. The Scorecard Execution Roadmap highlights various steps to be executed for successfully implementing a business scorecard. These steps are implemented with the help of a set of forms in a timely manner.

The execution of a business scorecard begins with establishing a baseline, or the present state. It is quite possible that initially all BPIn measurements may not be available in an organization. Under these circumstances, the present state is determined using indirect measurements as identified on a form created for each BPIn measurement separately. These measurements are graphically displayed by thermometers, where the user can see actual performance, as well as become aware of status of the BPIn measurements through color zones. Various BPIn measurements are evaluated, and opportunities are identified and prioritized for actions in the short or long terms. In case, an organization needs to determine its sigma level, where Six Sigma is being implemented, one can use the BPIn to DPU conversion, determine the DPMO, and estimate the sigma level.

Once the opportunities have been identified and recognized by the executive team, corporate vision, beliefs, goals, and initiatives are evaluated or created. In the absence of corporate vision and beliefs, it becomes difficult to create a corporate culture that is aligned with the leadership defined business objectives. Key business processes are identified, such as engineering, production or operation, sales, quality, and the like. These processes are mapped to the BPIn measurements. For each measurement, action plan is developed to achieve desired results or sustain the desired performance. The action items identify clear responsibility, speed of completion, and critical needs ensuring implementation. Finally, a monthly BPIn tracker is given for monitoring business performance.

The monthly tracker shows color zones to draw facilitate appropriate response. The "red" zone represents fundamental problems in an organization, and strategic changes are required. The "yellow"" represents need for improvement, thus operational changes are required. Finally, the "green" represents areas of excellence. If an organization is operating in the green zone, the leadership must focus on becoming the best in class through innovation and excellence in leadership, strategic, and operational areas.

As shown in the Scorecard Execution Roadmap, Six Sigma and Lean-type tools can be used to aggressively achieve excellence in operations. Brinnovation, a system developed by the author for institutionalizing innovation for achieving growth. The 4P model of process management has been developed by the author to sustain excellence. In case of Six Sigma, the focus must be on its intent to accelerate improvement, and the DMAIC methodology. The statistical thinking is more important than using many of the advanced statistical tools. The Brinnovation highlights need for understanding innovation before starting to measure innovation. The basic building block of innovation is a "networked individual" with unlimited potential for innovation. The Brinnovation methodology emphasizes commitment to innovation throughout the organization, and promoting intellectual engagement of all employees. Thus achieving excellence in idea management becomes an imperative in the current economic environment.

As to strategic changes, when in the red zone, the organization must look into its intent, purpose, product and services, culture, leadership style and commitment, its constraints, systems, competition, and industry trends. The comprehensive self-assessment will highlight key areas for strategic changes which must be used to develop new strategic direction, and steering the organization to operational excellence and superior financial results for all stakeholders.

The chapter highlights need for developing a strategy that can be executed in an organization effectively by focusing on value-added business processes, and clear responsibilities. The objective is to make a strategy much more successful than

10 percent, as generally reported. Strategy maps are complex, difficult to understand, and strategy focused. Instead, the execution map is logical, easy to understand, implement, and linked to the fundamental strategy of an organization.

TAKE AWAY

1. Strategy alone is not good enough, and does not do much for the organization.
2. Strategy must be developed for execution by knowing current capability, constraints, and resources.
3. The SFE requires a clear knowledge of an organization, and the execution map, which consists of measurement, improvement, innovation, and management.
4. The tools for the execution map include the Six Sigma Business Scorecard, Six Sigma, Innovation, and the 4P model.
5. Organizations must establish its vision, beliefs, and goals to synergize its energies
6. The execution map has been provided for consistent, and efficient implementation of the fundamental business strategy of sustained profitable growth.

ADAPTING THE SIX SIGMA BUSINESS SCORECARD TO SMALL BUSINESSES

Many small businesses perceive implementation of a measurement system as a distraction. The perception is that such measurement systems or fancy business scorecards are designed for large or public corporations. Although such claims have some merit, this understanding should not become a barrier to implementing a business scorecard. Small businesses can make adjustments to the measurement system based on the functions and size of the organization. Balancing profitability and growth is a common objective irrespective of the business's size.

SIX SIGMA BUSINESS SCORECARD AND BUSINESS SIZE

A business starts with one person before it starts growing. For a start-up business, the Six Sigma Business Scorecard categories can be used as guidelines to consider while managing and growing the business. Established small businesses can adapt the Six Sigma Business Scorecard according to the organizational functions. The Six Sigma Business Scorecard categories can be reviewed for implementation in a start-up or a small business as follows:

SIX SIGMA BUSINESS SCORECARD CATEGORIES	APPLICABILITY TO A START-UP BUSINESS OF 1 TO 25 EMPLOYEES	APPLICABILITY TO AN ESTABLISHED SMALL BUSINESS OF 26 TO 75 EMPLOYEES	APPLICABILITY TO A MEDIUM-SIZE BUSINESS OF 76 OR MORE EMPLOYEES
Leadership and profitability	Yes, measurable	Yes, measurable	Yes, measurable
Management and improvement	Yes, guidelines	Yes, guidelines	Yes, measurable
Employees and innovation	Yes, guidelines	Yes, guidelines	Yes, measurable
Purchasing and suppliers' management	Yes, measurable	Yes, measurable	Yes, measurable
Operational excellence	Yes, measurable	Yes, measurable	Yes, measurable
Sales and distribution	Yes, measurable	Yes, measurable	Yes, measurable
Service and growth	Yes, guidelines	Yes, measurable	Yes, measurable

Many small businesses consider the cost of implementing a comprehensive system to be unjustifiable because of the effort and resources required, especially for nonproduction activities. However, small businesses cannot ignore the value of good measurements. The Six Sigma Business Scorecard enables businesses of all sizes to review business performance frequently and assess it against the industry practices. More important, the Six Sigma Business Scorecard can drive the improvement and help keep a business from being dependent on only a few important customers. The Six Sigma Business Scorecard does require a firm commitment for its implementation.

To balance available resources and the requirements for implementing the Six Sigma Business Scorecard successfully, a small business must look significantly at more objective measures of performance rather than subjective measures of recognition. For this reason, three measurements were removed from the scorecard, and weightings were modified to reflect the new organization. The revised scorecard for small businesses is shown in Figure 10-1. Once the BPIn is calculated for small businesses, Figure 10-2 converts the BPIn to DPU for ease of use.

Measurements	Applicability to Small Businesses	Importance A	Initial Performance Guidelines B	% Score C	Weighted Score (AxC/100) D
1. Employee Recognition (% of employees)	Guidelines	NA	0.2% - 25		
			0.5% - 50		
			2% - 75		
			>5% - 100		
2. Profitability	Measurable	30	2% - 50		
			4% - 60		
			8% - 80		
			>12% - 100		
3. Rate of Improvement in Process Performance	Guidelines	NA	<20% - 50		
			30% - 60		
			40% - 80		
			>50% - 100		
4. Recommendations per Employee	Guidelines	NA	0.5/Emp - 50		
			1/Emp - 60		
			2/Emp - 70		
			>5/Emp - 100		
5. Total Spend/Sales	Measurable	10	>60% - 30		
			45% - 50		
			35% - 75		
			<25% - 100		
6. Suppliers' Defect Rate	Measurable	10	3σ - 25		
			4σ - 50		
			5σ - 75		
			6σ – 100		
7. Operational Cycle Time Variance	Measurable	10	>50% - 25		
			40% - 50		
			25% - 75		
			<10% - 100		
8. Operational Sigma	Measurable	10	<3σ - 25		
			4σ - 50		
			5σ - 75		
			>6σ – 100		
9. New Business/ Total Sales	Measurable	10	20% - 25		
			30% - 50		
			40% - 75		
			50% - 100		
10. Customer Satisfaction	Measurable	20	80% - 60		
			85% - 70		
			90% - 80		
			100% - 90		
Total (BPln)					
DPU (-ln(BPln/100)					
# of Executives (president and staff)					
DPMO					
Sigma (from table)					

FIGURE 10-1.   Six Sigma Business Scorecard for small businesses.

BPIn	DPU	BPIn	DPU
10	2.302585	55	0.597837
15	1.89712	60	0.510826
20	1.609438	65	0.430783
25	1.386294	70	0.356675
30	1.203973	75	0.287682
35	1.049822	80	0.223144
40	0.916291	85	0.162519
45	0.798508	90	0.105361
50	0.693147	95	0.051293

FIGURE 10-2.   BPIn-to-DPU conversion table.

In implementing the Six Sigma Business Scorecard, small businesses' main concern is data collection and analysis and managing improvement. Often, small businesses do not clearly assign responsibility for process improvement because they don't have the infrastructure or enough people to go around. Small businesses have some advantages over their larger counterparts in that sorting out questionable material and sending on good material are much easier for them. As the businesses grow, these activities matter more and more. The Six Sigma Business Scorecard can be a valuable asset for small businesses that see opportunity for growth.

Once the scorecard is established for a small business, the process of using the information is the same. An established methodology such as Six Sigma can then be used to address problems. The advantage of implementing Six Sigma is that many large companies are implementing it, and they demand their suppliers do so as well. Similarly, implementing the Six Sigma Business Scorecard will force the company to improve its profitability and growth, because the business scorecard improves visibility of problems in an organization.

IMPLEMENTING IN A SMALL BUSINESS

The following steps can be used as guidelines in implementing the Six Sigma Business Scorecard in a small business:

1. Assign a person to be the champion for implementing and maintaining the scorecard measurements.

2. Establish a data collection mechanism and a database in which to enter the data.

3. Define the analysis needs and reporting procedures.

4. Review the data and take necessary action in a timely fashion.

5. Schedule a monthly review to go over the Six Sigma Business Scorecard.

6. Review the measurements and correlate them with profitability and growth.

7. Take appropriate actions to improve profitability and growth.

Diversity is another aspect of the business that must be managed for improved performance. Many small businesses have a large proportion of bilingual workers. Under those circumstances, the leadership must work to ensure that all employees understand their roles clearly and execute their responsibilities with excellence. Communicating the company's vision, beliefs, and goals helps to establish a common understanding of the business scorecard process.

As someone once said, a small company does everything a large corporation does, except the small company does everything a little faster and cheaper. The responsiveness of small companies is what enables them to grow in niche markets. Losing responsiveness and increased costs are two main constraints on business growth identified by small company leaders. Especially in small businesses, implementing the Six Sigma Business Scorecard must not become a burden on company resources.

Since small companies have a simplified business organization, the Six Sigma Business Scorecard can be implemented using computer resources. Automating the data analysis and reporting mechanisms is much easier to do in a small business than in a large corporation. Integrating the Six Sigma Business Scorecard with the current business management system, such as ISO 9000, is also key. The review of the Six Sigma Business Scorecard can be integrated with management review meetings, and the analysis of the scorecard data aligns with the data analysis requirements of the ISO 9000 system.

Small business owners often believe that they already know everything about their business and that they do not need additional measurement systems to tell them about the business performance. Their main measurement is the profitability at the end of the day. To some extent they are right. However, if the business is set to grow and the leadership is focused on diversifying products and services, the Six Sigma Business Scorecard can help in maintaining the information flow and keeping an eye on profitability—the small business leadership's main objectives.

Ultimately, the Six Sigma Business Scorecard was designed to manage the business in its entirety and balance its multiple dimensions. Monitoring all aspects of the business can become increasingly difficult for the leadership. This is especially true for small businesses, as the business leaders already perform multiple tasks, such as technical consulting, sales, and financial management. In addition, the executive and professional staff resources are limited in small businesses. Using the Six Sigma Business Scorecard to its fullest potential can maximize people power and profitability. It can help small businesses to achieve better profitability and growth objectives as well.

TAKE AWAY

1. Small businesses can make adjustments to the measurement system based upon the functions and the size of the organization.

2. Balancing profitability and growth is a common objective irrespective of the business's size.

3. A small company does everything a large corporation does, except the small company can do everything a little faster.

MONITORING PERFORMANCE USING THE SIX SIGMA BUSINESS SCORECARD

Simply implementing the Six Sigma Business Scorecard is not enough to improve the business's performance. The Six Sigma Business Scorecard must be implemented *interactively*; that is, the management and employees must be involved in reviewing the information represented by the scorecard and taking actions based on the interpretation of the data. They need to look at levels, trends, and rate of improvement. In today's competitive environment, the rate of improvement must match the combined rate of change in customer expectations, market conditions, and society.

Remember, in order to be profitable and achieve growth, a company needs to monitor employee recognition, rate of improvement, and innovation in addition to the standard performance measurements of costs, quality, and cycle time. The objective of the review process is to ensure these indicators demonstrate the corporate performance as planned.

MANAGEMENT REVIEW

One of the weakest links in business management is management's review of performance. In a typical review, the performance data are presented, most of the attendees wait for their turn to speak and agree about what is presented, and the meeting ends with consensus about the data presented. In visualizing the process before the meeting, the main effort goes into preparing the presentation material for the meeting—not necessarily understanding performance. As a result, the focus of the meeting is the presentation instead of the performance. A healthy critical analysis of performance, an understanding of the variation in performance, and actions to improve further are often overlooked. A stronger review process must be established and practiced for a business to benefit from the management review.

The purpose of the management review is for leadership collectively to understand a company's performance, look into opportunities for further improvement, identify conflicting priorities among various departments, and assess the strategies being implemented for effectiveness and operational execution for expected results. An effective management review must be planned with a clear agenda. In a multidivision company, the frequency of executive management review must be coordinated with the reviews of various facilities so that the data are aggregated from bottom up to the corporate level. Figure 11-1 shows the corporate review cycle. The cycle includes a monthly review of operations at the process, division, and corporate levels in addition to a conventional quarterly review of the financial measures in detail. The monthly review will drive a better quarterly corporate performance.

To prevent these reviews from becoming a trivial exercise, the review guidelines at each level must be clearly established and the goals of the Business Performance Index and Six Sigma Business Scorecard must be communicated. Rather than review performance at a point in time, these meetings must review performance against the sliding goal line on the trend charts. They must assess the improvement from the previous review as well as project performance in the future. The interaction of the various departments must be reviewed for improving synergy and mitigating adversities.

To review performance using the Six Sigma Business Scorecard, the agenda must include examining each depart-

FIGURE 11-1. Corporate performance review cycle.

ment for cost, rate of improvement, and timeliness. The status of action items from the previous meeting must also be on the agenda. The main objective of the review is to ensure that performance goals are achieved or exceeded. The suggested agenda items are as follows:

1. Status and effectiveness of action items from the previous meeting

2. Review of measurements by elements

   2.1 Leadership and profitability

   2.2 Management and improvement

   2.3 Employee innovation

   2.4 Purchasing and supplier management

   2.5 Operational execution

   2.6 Sales and distribution

   2.7 Service and growth

3. Division or corporate Sigma level

4. Action items for continual dramatic improvement

5. Suitability of the Six Sigma Business Scorecard

During the review, attendees present information about their designated category or assigned measurements for level, trend, rate of improvement, and Sigma. They then analyze the data for statistically significant changes by identifying nonrandom patterns. If unacceptable patterns exist, the team at the meeting identifies an action to remedy the systemic problem.

LEADERSHIP PERFORMANCE REVIEW

Each stakeholder is accountable to someone in a corporation. The CEO is directed by the board of directors and the board chair. However, a CEO is hardly graded for a process performance; instead, the CEO is graded for results. CEOs tend to be evaluated for their interaction with media instead of their interaction with employees.

LEADERSHIP REVIEW

Traditionally, leadership's performance has been measured in terms of outcomes that are sometimes difficult to predict. However, decisions about the leadership's performance are made based on the short-term performance of the business or lack of it. For the leadership performance to be evaluated consistently, the leadership must be evaluated as a process by stakeholders, including employees.

The leadership evaluation process is shown in Figure 11-2. The inputs into the leadership evaluation process consist of information, executives, policies, and tools. The information portion of the process should include the business scorecard measurements as well as any other information that would help identify opportunities for improvement or provide feedback about leadership's performance. The executives referred to in the process are the CEO and staff members—the people who

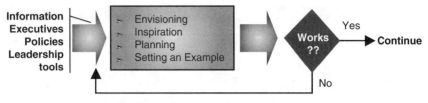

FIGURE 11-2.   Leadership evaluation process.

serve as the resources to realize the leadership's vision. The CEO must be evaluated for the quality of his or her team and its performance. The policies include the guidelines and mandatory requirements that the leadership must meet, such as government and trade regulations, internal procedures for various tasks the CEO performs, or the goal-setting process that allows for dramatic improvement. The leadership tools may include mechanical tools, informational tools, and training tools to enhance the leadership's performance.

Stakeholders' feedback must be received and taken into account in the leadership compensation plan. Leadership must seek employees' feedback on how well the leadership demonstrates the same characteristics expected of employees. Similar feedback must also be solicited from other stakeholders, such as shareholders, suppliers, and the board of directors. The leadership must be bold enough to accept and make use of the critical feedback for improving its own performance. This can be accomplished through a standard set of questions that address various aspects of the leadership. Figure 11-3 provides a sample Leadership Performance Review that can be customized as appropriate.

Leadership evaluation is critical in sustaining the success of a company. A company will only be as successful as its leader. Leadership is an evolving process that requires monumental on-the-job training. Many workplaces perpetuate the misunderstanding that the leader is the smartest and most successful person in the company. That may have been true in the past. However, in each new assignment, the leadership process must be adjusted to fit the company's goals, and that adjustment requires a lot of input because the leadership's action may affect

Leadership Performance Review	Employees' Feedback
Envisioning	
Always thinking; analyzing competitors, new business strategies, new work methods, and the future	
Not afraid of experimenting, striking out in new directions	
Buoyant and building a sense of excitement and adventure in employees	
Takes risks and pursues novel approaches and encourages creativity in others	
Demonstrates mastery over crucial skills and is eager to share the knowledge	
Ever-present willingness to learn new things, to explore new areas, to test concepts, and is not afraid of failure	
An excellent reader and builds on others' ideas	
Inspiring	
Candid, honest, and objective	
Quick to recognize and thank others	
Understands his or her success is a team effort and a result of good fortune	
Listens well in order to learn from others	
Cooperative and enjoys working with others	
Perceptive of thoughts and fears of others	
Shows exuberance – a childlike curiosity inspiring others	
Planning	
Keeps his desk, agenda, and mind open and clear	
Prioritizes tasks and devotes necessary effort	
Establishes his own best-in-class benchmarks	
A good strategist who systematically deploys available resources for best results	
Action-oriented and plans to make things happen	
Plans to continually do better and expects others to follow	
Plans to get involved as needed to achieve results	
Practicing (Exemplary)	
Demonstrates exemplary dedication	
Looks at all aspects of an issue before making a decision	
Practices and encourages fairness	
Tenacious and keeps moving forward toward goal that appears elusive to others	
Lets others bask in limelight	
Builds and maintains comfortable, goal-oriented atmosphere at work	
Strives for excellence in all things. Does the best job possible and inspires others to do the same.	

FIGURE 11-3. Leadership performance review elements.

everyone in the company as well as those whom the company does business with. That's why the people who will be affected by the leadership process must be enlisted in achieving common goals. Leadership responsibility is given; however, the authority is earned. In the absence of proper leadership evaluation, many successful leaders fail in their assignments.

EMPLOYEES' PERFORMANCE REVIEW

Unlike the leadership performance review, the employees' performance review process is more robust and measurable. Because the number of employees can be magnitudes larger than the number in the leadership, a standardized performance evaluation process is critical to achieving the leadership vision. Without an effective employee performance review process, consistently achieving the business objectives is a daunting task.

Deming has challenged rating employees based on a performance review. According to Deming, variation in employees' performance is mainly due to the system of which they are a part. Employees perform what they are expected to perform. The perception among employees is that employee ratings are used unfairly in tough times to justify eliminating positions and in good times for providing growth opportunities. The employee performance review is not necessarily a human resources management issue; instead, it is a process that must be aligned strategically to achieve the business objectives. It must be used dynamically to achieve business goals. Figure 11-4 shows how

Employee Performance Review
Excellence Expectations
Leadership Practices
Value Added (including Six Sigma projects)
Innovation
Demonstrated Excellence (Measurable Value Added)
Innovation (including breakthrough solutions)
Direct Value-Added
Leadership Practices
Areas for Improvement
Behaviors
Skills
Growth Plans
Training
New Assignments
Incentives for Extraordinary Performance
Value Plans (Including Six Sigma Projects)
Plans for Innovation
Thought Leadership
New Areas of Interest

FIGURE 11-4. Employee performance evaluation elements.

employee reviews can be linked to ongoing employee performance.

Formal annual performance reviews aren't enough to help
employees improve performance. Rather, they must be accom-
panied by more frequent communication about performance,
such as weekly reports or monthly operations reviews. These
reports and reviews must have a format that links to the annual performance reviews, which are then directly linked to a
company's strategic initiatives. Performance recognition is an
ongoing process. This notion must be reflected by leadership
through behaviors, body language, decisions, and actions.
Financial incentives are a prime motivator of performance
improvement. But recognition coupled with financial incentives that are based on a fair distribution mechanism can lead
to dramatic improvement in performance.

Innovation is a critical aspect of the Six Sigma Business
Scorecard. Innovation must be promoted at every level and at
every opportunity. Leadership must plan to help employees
develop new skills and expand their roles to help the company
grow. Employees need to learn and practice leadership skills.
They must also be encouraged to expand their intellectual
involvement and personal growth. Such encouragement creates
growth opportunities for the company and leadership opportunities for employees.

Promoting innovation is a challenging task, and utilizing the
brain of every employee must become a corporate strategy.
Acquiring the mental involvement of employees is daunting,
because it requires employees' personal initiative. That normally
happens when the employees are self-motivated (a leadership
skill), or when there is a significant incentive for recognition or
reward, a challenge to create something new, or a desirable performance expectation aligned with the corporate initiative. The
leadership must create challenging opportunities with recognition and rewards in order to motivate employees to innovate.

MANAGEMENT PERFORMANCE REVIEW

The main objective of the management must change from its
traditional role of managing operations to one of achieving the

necessary rate of improvement in their areas. In supervising operations, managers must ensure that their employees understand the area or department objectives, have necessary skills and tools to achieve them, and use those tools efficiently and effectively. In addition, managers must make sure that departmental processes are monitored effectively and that performance is reported accurately. In reporting performance, managers must track levels and trends in DPU and DPMO as well as Sigma.

When success in the execution of corporate strategies is considered, management performance review is generally very weak. This is so often because managers either do not understand the corporate objectives or are not given enough time or resources to understand them well. Sometimes, the leadership team simply dumps its goals on the management staff, or worse yet, sometimes managers are unaware of the corporate goals. The management team is the "execution" team, and its members must be fully enlisted in developing realistic goals based on facts, as well as in developing plans to achieve them. They are the real drivers of the dramatic improvement. While the leadership may design goals and objectives, management must own the objectives in order to achieve them.

The management performance review must incorporate assessments in the key areas of rate of improvement, breakthrough solutions, direct value creation (including Six Sigma projects), and leadership practices (see Figure 11-5). According to the Six Sigma Business Scorecard, the main emphasis on the management cadre is to achieve the desired rate of improvement, generate savings, and recognize employees for achieving the desired improvement and savings. To achieve a rate of improvement of 50 to 70 percent annually, management must involve all employees intellectually as well as have teamwork, innovation, superior resource management, and periodic review of the departmental performance.

The performance evaluation for rate of improvement, breakthrough solutions, direct value creation, and leadership practices must be performed against clearly established expectations for improvement. At the management level, excellence must be ingrained through training and communication. Managers must apply superior work ethics, inspire innovation

Management Performance Review
Improvement Expectations
Rate of Improvement
Breakthrough Solutions
Direct Value Creation (including Six Sigma projects)
Leadership Practices
Demonstrated Improvement (Measurable Value Added)
Rate of Improvement
Breakthrough Solutions
Direct Value Created
Leadership Practices
Areas for Improvement
Improvement Goals Not Met
Resource Management
Growth Plans
Leadership Training
New Areas of Interests
Incentives for Group's Superior Performance
Value Action Plans (Including Six Sigma Projects)
Plans for Innovation
Thought Leadership

FIGURE 11-5. Management performance evaluation elements.

in their departments through teamwork, maintain a sense of urgency, and demand value creation.

In many companies, management has said, "Let's work smarter, not harder." In this competitive environment, however, we must work both smart and hard to the best of our abilities. Being smart without the desire to put that smartness to good use is a waste of valuable corporate resources. In many companies, smartness is used to manipulate the measurement system instead of to achieve greater improvement. Such "smart" tactics should never be permitted.

Managers, therefore, must be expected to deliver results and be assessed for demonstrating results and positive behaviors. Again, the goal of performance evaluations is to strengthen the management process, not to punish weak managers. At the management level, failure is not personal—failure is the system's failure, since management personnel have been successful in previous years. In the case of poor performance, questioning a manager's individual integrity, desire, and dedication is counterproductive. Instead, the focus of the effective feedback should be on aligning management goals with the leadership and effectively applying management skills to achieve business objectives. Continual improvement at an aggressive rate must be management's mantra, and subconscious actions must communicate management's performance.

COMPENSATION FOR PERFORMANCE

Successful implementation of Six Sigma combines expectations for dramatic improvement with recognition and reward. At Motorola, teams achieving extraordinary improvement and cost reductions shared the savings based on formulas designed to ensure corporate profitability. At General Electric, senior-level executives are expected to perpetuate the Six Sigma philosophy and methodology. For Six Sigma to be effective, the organizational structure must be revised to inspire, stretch goals, encourage innovation, plan dramatic improvement, and achieve significant savings.

The leadership and management must ensure savings through effective planning and superior execution. Incentives, such as sharing in any savings realized and personal growth opportunities, must be linked with commitment to corporate strategies, measurement using a Six Sigma Business Scorecard–like system, and profitability improvement. The revised organizational structure requires the business to identify opportunities for improvement, develop a sound corporate strategy, and train the necessary human resources, including management. The organization must have a Chief Growth Officer (CGO) to drive growth initiatives, a Scorecard Manager to monitor performance, and Project Managers to realize project benefits.

While devising compensation plans for employees to reward them for improvement, there are several things to keep in mind. Most important, the improvement must be real without any manipulation of numbers or opportunities. The improvement and savings must be visible and measurable without any uncertainty. If any doubts about improvement and savings exist, compensation must wait for the next cycle when the real improvement and benefits are realized.

COMMUNICATION WITH THE COMMUNITY

Ensuring continual commitment to the Six Sigma initiative and the Six Sigma Business Scorecard requires setting appropriate expectations within the entire community that a business touches. All customers, stakeholders, employees, and suppliers must understand the company's commitment to the Six Sigma initiative. Businesses must routinely communicate a consistent message to all players, including employees, suppliers, customers, and stakeholders. They may want to regularly report on the status of institutionalization of Six Sigma, success of projects, resultant savings, and alignment with business initiatives.

An important aspect of the communication is to ensure that both the leadership and the stakeholders do not focus on the numerical goals of Six Sigma instead of on the institutionalization of the methodology. The institutionalization must include a common understanding of Six Sigma, the leadership's vision, companywide project selection, and personnel growth.

In the case of Motorola, the expectation, communicated by the leadership, was that the company had a 5-year plan to implement Six Sigma. As a result, the general perception was that the company would achieve a defect rate of 3.4 parts per million throughout the corporation. This, however, was an impossible task due to changing customer requirements and new products. Even experts predicted that, based on a statistical analysis, achieving a defect rate of 3.4 parts per million was impossible. However, Motorola's plan focused on institutionalizing the Six Sigma methodology throughout the corporation and across the supply chain. This goal was mainly accomplished as planned; however, the journey to achieve a defect rate of 3.4 parts per million continued beyond 1992.

ANNUAL REVIEW

Besides having annual goals for Six Sigma implementation and using the Six Sigma Business Scorecard, successful implementation requires an annual review for suitability, adequacy, and effectiveness of the corporate strategies, including Six Sigma. Finally, an independent review by a third party will ensure the integrity of the Six Sigma Business Scorecard.

TAKE AWAY

1. The Business Scorecard must be implemented interactively—the management and employees must be involved in reviews and actions.

2. The rate of improvement must exceed aggregate change in customer expectations, market conditions, and cost of inflation.

3. The leadership must be evaluated as a process by stakeholders, including employees.

4. Leadership must seek employees' feedback on how well the leadership demonstrates the same characteristics expected of employees. In the absence of proper leadership evaluation, many successful leaders fail in their assignments.

5. Promoting innovation is a challenging task and maximizing use of human capital must become a corporate strategy.

6. The Business Scorecard emphasizes the role of managers to lead the aggressive rate of improvement, generate savings, and recognize employees.

7. The improvement and savings must be visible and measurable without any uncertainty.

PERFORMANCE, PROFITABILITY, AND STANDARDS

The Six Sigma Business Scorecard and the Business Performance Index (BPIn) establish a system of key measurements that can help businesses maximize their profitability. Utilizing the Six Sigma Business Scorecard for profitability requires an understanding the common measurements used in the business world today. The *Dow Jones Industrial Average* (the *Dow*) is a well-known business performance index that is influenced by the listed companies' performance as well as the perception of its shareholders. In reality, the Dow represents the performance of sample companies that represent the U.S. stock market.

DOW JONES INDUSTRIAL AVERAGE COMPANIES

Thirty companies listed in the Dow Jones Index, also known as the Dow 30, are shown in Figure 12-1. The companies represent 10 economic sectors, including the industrial, technology, telecom, financial, health care, and consumer sectors. Operationally, these companies can be grouped into fewer sectors, such as manufacturing (18 companies), service (4 companies), software (1 company), distribution (4 companies), and financial (3 companies).

Company	Sector	Company	Sector
Alcoa	Basic materials	Intel	Technology
American Express	Financial	IBM	Technology
AT&T	Telecom	International Paper	Basic materials
Boeing	Industrial	Johnson & Johnson	Health care
Caterpillar	Industrial	J.P. Morgan Chase	Financial
Citigroup	Financial	McDonalds	Consumer, cyclical
Coca Cola	Consumer, noncyclical	Merck & Co.	Health care
Dupont	Basic materials	Microsoft Corp.	Technology
Eastman Kodak	Consumer, cyclical	3M	Industrial
Exxon	Energy	Phillip Morris	Consumer, noncyclical
General Electric	Industrial	Proctor & Gamble	Consumer, noncyclical
General Motors	Consumer, cyclical	SBC Communication	Telecom
Hewlett-Packard	Technology	United Technology	Industrial
Home Depot	Consumer, cyclical	Walt Disney	Consumer, cyclical
Honeywell International	Industrial	Wal-Mart	Consumer, cyclical

FIGURE 12-1.   Dow Jones Index companies.

Companies in different operational sectors have some assignable variations in their financial statements that reflect variations in their business operations. For example, distribution companies have a relatively high inventory cost, while companies in the financial sector have practically no inventory. In addition, because companies such as IBM and GE have both manufacturing and financial operations, their income statements and balance sheets reflect that dual nature.

PERFORMANCE AND PROFITABILITY

The purpose of analyzing the Dow 30 data is to understand the relationships between profitability and various reported measurements, and to understand variances in key measurements. The key measurements include

- Profits
- Sales
- Cost of goods sold
- Research and development
- Inventory
- Sales, general, and administration expenses

- Assets
- Long-term debt
- Shareholders' equity
- Plant and equipment
- Retained earnings

These measurements reflect cost and revenue streams, key processes, and the main financial measurements that are used to report a business's performance. Using regression analysis, the effectiveness of these measurements was reviewed. Figure 12-2 summarizes these measurements for the Dow 30 companies.

Exceptions in every measurement exist due to the nature of the business; however, the business model remains the same. For example, several large financial houses, such as Citigroup and J. P. Morgan Chase, show minimal expenses in the plant and equipment category. Microsoft has no debt. Several companies do not report itemized research and development expenses, even though almost every business conducts some research and development for new products or services. The Dow 30 measurements demonstrate that the average *cost of goods sold* (COGS) is about 57 percent of sales, the *sales, general, and administration* (SGA) expenses account for 15 percent of total sales, profitability is about 8 percent, and other expenses account for about 20 percent. The other expenses may include interest to serve loans, accrued employee benefits, and retained earnings.

Measurement	Dow 30 Average	Low	High	% of Sales
Sales	$58,114	$13,234	$217,799	100
Cost of Goods Sold	33,044	3,455	171,562	57
R&D	2,045	203	5,290	4
Inventory	4,928	105	22,614	8
SGA Expenses	8,977	1,276	32,173	15
Profits	4,350	−1,204	15,320	8
Long-Term Debt	16,015	0	121,631	28
Plant and Equipment	19,226	1,903	89,602	33
Shareholders' Equity	25,167	2,894	81,247	43
Retained Earnings	22,505	−3,484	95,718	39
Total Assets	$133,635	$13,362	$1,051,450	230

FIGURE 12-2. Summary of key measures (in millions).

Analyzing the data shows that profitability has a strong relationship with the cost of goods sold, SGA expenses, and sales, given the complexity of business dealings. Therefore, performance measurements must include gathering data on the cost of goods sold and operational expenses and setting goals to reduce them. To establish a causal relationship between the performance measurements and profitability, correlations that capture the relative rankings of various measurements in order of significance were reviewed, as shown in Figure 12-3. A strong or weak correlation between two factors does not ensure a causal relationship, but it does indicate that there may be one. Figure 12-4 shows that profits are strongly correlated with a combination of the three key factors, i.e., sales, cost of goods sold, and SGA expenses. However, individually these three measurements are mildly correlated in the following order:

1. Sales
2. SGA expenses
3. Cost of goods sold

To improve profitability, a business must increase sales, reduce the SGA expenses, and reduce the cost of goods sold (in this order). Of course, there is a tradeoff between the three variables, so a balance must be found to maximize profitability. As Figure 12-4 illustrates, about 20 percent of the margins are accounted for in indirect expenses, and profitability correlates strongly with

Measurements	Correlation with Profits	Correlation with Growth, 3 Years' Sales
Plant and Equipment	0.6936	0.2365
Long-Term Debt	0.5891	0.1795
Total Assets	0.4623	0.1681
Shareholders Equity	0.7861	0.1514
R&D	0.2943	0.1494
Profits	1.0000	0.1125
Retained Earnings	0.8346	0.0690
Sales, Cost, SGA	0.8748	(0.1239)
Sales	0.6048	(0.1586)
SGA Expenses	0.4829	(0.2008)
Inventory	0.4219	(0.1315)
Cost of Goods Sold	0.2434	(0.2328)

FIGURE 12-3.   Financial breakdowns for Dow 30 companies.

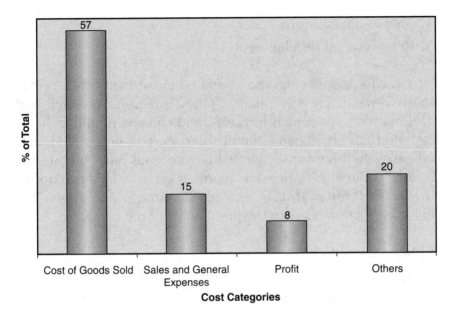

FIGURE 12-4.   Profits and sales growth versus business performance measures.

factors such as shareholders' equity, retained earnings, and plant and equipment expenses. However, the business needs to understand the causal relationships on a case-by-case basis and then establish internal goals and targets. As the profitability equation indicates, profitability is a dependent variable, and cost and expenses are independent variables, while the equity and retained earnings are necessary for growth instead of profitability.

MEASUREMENTS FOR GROWTH AND PROFITABILITY

Reviewing the relationship between 3-year sales growth and various business performance variables, we see that the correlations are insignificant. However, a relative ranking identifies the following factors in their order of significance:

1. Plant and equipment
2. Long-term debt
3. Total assets

4. Shareholders' equity

5. Research and development

Growth depends on innovation and investment, while profitability depends on performance. Growth requires research and development, investment in plants, and equipment and technology. Profitability depends upon sales, cost of goods sold, and expenses. In other words, profitability depends more on internal factors, while growth depends more on external factors. *Growth is realized through strategy and profits through execution.* The following factors relate to the profitability of the company:

1. Sales-cost-SGA

2. Sales

3. SGA expenses

4. Inventory

5. Cost of goods sold

To maximize growth, the leadership of a business must drive innovation by encouraging employee participation and investing internal and external financial resources. To maximize profitability, management must optimize the profitability equation by increasing sales, reducing SGA expenses, and reducing its inventory of resources. Cost of goods sold has a strong relationship (88 percent) with the inventory.

To achieve both profitability and growth, various factors must be managed at the process level. However, profitability and growth are two distinct but related events that must be managed consciously and can be optimized for best results.

BALANCING THE SIX SIGMA BUSINESS SCORECARD FOR GROWTH AND PROFITABILITY

The Six Sigma Business Scorecard incorporates factors that lead to both profitability and growth. As Figure 12-5 shows,

IMPROVEMENT

INNOVATION

FIGURE 12-5.   Six Sigma Business Scorecard.

the scorecard has seven categories. Some categories are growth-related, while others are profit-focused. The following table shows the breakdown of categories according to profitability and growth:

CATEGORIES	PROFIT/GROWTH
Leadership	Growth
Sales and Distribution	Profit
Purchasing	Profit
Operational Execution	Profit
Rate of Improvement	Profit/Growth
Employee Innovation	Growth
Service and Growth	Growth

To improve profitability, key measurements are highlighted to maintain visibility of the profitability. These measurements are Purchasing, Operational Execution, and Sales and Distribution. Other measurements highlighted to drive growth include Leadership, Employee Innovation, and Service and Growth. Leadership establishes a process for recognizing employee excellence and extraordinary effort; employees innovate new solutions, products, or services; and customer service understands customers and their present and future needs. Accordingly, the leadership drives the growth, and the executive team manages the profitability. One must establish drivers for both profitability and growth. Without profitability no growth can be sustained, and without growth no business can survive. Therefore, both growth and profitability must be managed with equal vigor.

Considering the pressures from the stock market for quarterly performance and high expectations, many companies have achieved significant results by maintaining profitability as well as growth. *Fortune* magazine's 2002 Global Most Admired Companies for Financial Soundness include ExxonMobil, Wal-Mart, Intel, Microsoft, Home Depot, General Electric, United Parcel Service, Walgreen's, and Citigroup. The leaders at these companies, including the boards of directors, are responsible for setting the right expectations in terms of the profitability and growth over a period of time, rather than maximizing the profits and stock value in the short term, which can bankrupt the company in the long term. Their communication with internal and external stakeholders establishes the desired expectations and builds confidence in the company for the long term.

BUSINESS PERFORMANCE INDEX (BPIn)

For the BPIn, measurements can be broken down into profit and growth categories, as shown:

MEASUREMENTS	PROFIT/GROWTH
Employee Recognition	Growth
Profitability	Profit
Rate of Improvement	Profit/Growth
Recommendations per Employee	Growth
Total Spending/Sales	Profit
Suppliers' Defect Rate	Profit
Operational Cycle Time	Profit
Operational Sigma	Profit
New Business	Growth
Customer Satisfaction	Growth

Additional measurements are monitored at the departmental and process levels. These include a larger number of measurements for leadership in order to promote innovation and profitability.

The BPIn measurements were identified based on analysis of several companies that have experienced losses, profits, and growth. These companies have implemented various measurements without considering them as elements of a scorecard. No one set of measurements or scorecard can fit all situations. Before a business establishes a performance measurement system, the leadership must determine what it expects to accomplish with a scorecard and what measurements will help achieve those goals. Even then, the performance measurement system must be dynamic and amended as needed to tune the performance in terms of its rate of growth and profitability.

Once the business determines what set of measurements it will use, the next question is, What are the established standards or acceptable performance levels? Each company must have its own acceptable standards based on its strategy and resources. Figure 12-6 shows a range of performance measurements that can be examined to establish a set of standards for a company. Any statistical analysis of variation *in performance*

Measurements	Category Abbreviation	Category Significance	Performance Guidelines
1. Employee Recognition (% of employees)	LNP	15	0.2% - 25
			0.5% - 50
			2% - 75
			>5% - 100
2. Profitability	LNP	15	2% - 50
			4% - 60
			8% - 80
			>12% - 100
3. Rate of Improvement in Process Performance	MAI	20	<20% - 50
			30% - 60
			40% - 80
			>50% - 100
4. Recommendations per Employee	EAI	10	0.5/Emp - 50
			1/Emp - 60
			2/Emp - 70
			>5/Emp - 100
5. Total Spend/Sales	PSM	5	>60% - 30
			45% - 50
			35% - 75
			<25% - 100
6. Suppliers' Defect Rate	PSM	5	3σ - 25
			4σ - 50
			5σ - 75
			6σ - 100
7. Operational Cycle Time Variance	OPE	5	>50% - 25
			40% - 50
			25% - 75
			<10% - 100
8. Operational Sigma	OPE	5	<3σ - 25
			4σ - 50
			5σ - 75
			>6σ - 100
9. New Business/Total Sales	SND	10	20% - 25
			30% - 50
			40% - 75
			50% - 100
10. Customer Satisfaction	SAG	10	80% - 60
			85% - 70
			90% - 80
			100% - 90

FIGURE 12-6. BPIn performance guidelines.

is practically impossible due to various economic sectors and the wide range created by the financial companies. Excluding such companies will not do justice to the analysis. Therefore, guidelines based on amount of improvement and acceptable performance should be compiled as standards that can be reviewed, revised, and approved for implementation.

MANAGING FOR PROFITABILITY

Managing profitability requires managing sales and costs. Therefore, leadership must focus on managing sales and managing costs explicitly.

MANAGING SALES

To maintain a specified level of profitability, the business must examine the components of its costs, expenses, and revenue. To increase revenue, sales strategies based on past performance, successes, and customer service must be developed. Getting repeat business is much easier than securing new customers. On the other hand, a business cannot depend indefinitely on just a few customers. To minimize the risk of excessive dependence on a customer, a process must be developed to acquire new customers.

The sales personnel are not the only individuals selling the products and services. Everyone in the company is selling directly or indirectly. The real sales pitch occurs through the performance once an order is secured. The revenue must be in line with the infrastructure and profitability objectives.

One of the challenges for growing revenue is that the sales price may be so low due to competition that the company does not make any profit. Yet the sales personnel still earn commissions on those sales as a standard practice. In such cases, how do those sales create profit for the company? They give headaches to the company leaders, stress to the employees, and, in the end, disappointment to the shareholders.

Clear guidelines for specific margins in the sales order must be established based on industry practices and cost structure. Leadership must decide whether to run the business at least as well as the rest of industry or better, or whether to sell on the lowest-price criterion. Any company focusing on selling the products or service at the lowest price alone will not survive long, because some competitor somewhere can always beat that price. Instead, sales guidelines must be based on their value to the customer as well as the total cost of the product or service. Usually, when the customer buys the cheapest product or part, there is a hidden additional cost revealed at the time of or after the sale.

Managing sales also involves strong customer relationship management. *Customer relationship management* does not mean selling the product or process at the cheapest price. Instead, it involves understanding customer needs well. Sales personnel should know what their customers need in terms of specifications as well how the product or service will be used and in what environment. Sales personnel can then develop a solution for the customer, providing a product or service that meets the customer's needs. Customers know that they need dependable suppliers that provide quality products or services. Suppliers must look for relationships that foster a real partnership that creates value. No business relationship can survive, no company can stay in business, and no industry can succeed based on the goal of promoting only the cheapest products or services.

Printed wiring boards (PWBs) in the U.S. electronics industry are a good example of a situation in which suppliers competed against one another based on price alone. As a result, the entire PWB industry suffered and eventually business migrated to global suppliers who had lower labor or material costs. This move violated the rule that suppliers must be in the vicinity to ensure effective communication on future product development, problem resolution, and success sharing. If the supplier-customer relationship is based on value and not on the price alone, the relationship could become cost-effective and long-lasting, irrespective of the distance.

REDUCING COSTS

The other side of managing profitability is to reduce costs. Costs include many components: inventory of production, nonproduction, or maintenance items; direct labor to produce goods or provide services; overhead costs; research and development; and so on. A business's leadership must understand the acceptable cost of doing business in the specific industry through external and internal benchmarking. The leadership must become knowledgeable about internal cost components. With this knowledge, leadership can set goals to reduce costs by a certain percentage every year to accommodate the cost of inflation and to maintain margins.

Everyone knows that customers continually expect better, faster, and cheaper products or services. However, many companies don't understand how much it costs them to produce and sell their products and services. Consequently, they lose profits because management has not done its job. As someone once said, the failure to plan for profits is a plan to fail miserably.

A company's management must manage the high-level business processes as efficiently as the operation-level processes. Profitability is an expected outcome of the business process. Management must maintain acceptable performance of inputs to the business and execution inside the business. Clear cost objectives must be in place and internalized to ensure they do not become hidden. Visibility of cost must be maintained so that excessive costs don't consume the entire business.

A business model and benchmarks are needed to keep a lid on expenses. The leadership must ensure that the sales expenses, general and administrative expenses, and other operating expenses remain within specified limits. They must track operating expenses as close to real time as possible, for such knowledge is the lifeline of the business. The leadership must watch these expenses on a monthly or quarterly basis and link its compensation to the components of profitability instead of to profitability alone.

In small, medium, and large businesses, the leadership often promises profits to shareholders at some later date that never materializes. That promise may have been a good goal; however, the leadership team's strategy and actions were not in line with those objectives. By the time shareholders realize that profitability objectives are going to be missed, it is usually too late to respond by cutting expenses and cost without jeopardizing the existence of the company.

In managing for profitability, the leadership must measure what matters. The measurements must be set for cost and sales components. Goals to achieve and goals to improve continually must be in place. With an effective leadership and superior management, the goals can be achieved and profitability can be sustained. Any business is either improving or worsening. There is no status quo to maintain. Survival is found in continually improving all aspects of the profitability equation; otherwise, the business is destined to die.

FACILITATING GROWTH

Business growth can come from repeat business or new business. Real business growth occurs when the business provides new products or services, thus acquiring new customers. To grow the business, the company must constantly monitor innovations that take place within the company and in the marketplace. External environmental factors must be monitored to drive internal innovation. Innovation can occur at any level, from a small component to a simple process. The objective is to make the product or perform the service better, faster, and less expensively tomorrow.

Generally, an organization is structured so that the technology officer drives internal research and development, sales is focused on obtaining more sales, and marketing explores new areas. These functional areas aim to excel in their span of control and in achieving results. However, because business growth objectives often are not clearly established, they cannot be communicated and achieved. Just as with managing profitability, someone must be directly responsible for managing growth—someone who can establish clear goals as well as a well-defined process and teamwork for achieving those growth objectives.

There is often a disconnect between what R&D has designed, what the market wants, and when the market wants it. Just as when someone throws darts in the dark, some ideas may hit the intended target, while others will miss entirely. If objectives can be established and a direction can be set, the growth objectives are more likely to be achieved consistently rather than sporadically. Growing companies develop an internal focus on growth, where synergy is created and the business runs as a well-oiled machine. Many other businesses that do not have this focus, however, struggle to grow.

INSPIRING FOR INNOVATION

Employees' intellectual participation is critical to growth. R&D is not the job of a few employees; on the contrary, every employee must be encouraged to innovate and perform his or her task profitably and creatively. Robert Galvin, former chairman of the board at Motorola, once said that he had 10,000 managers and 100,000 employees. He wanted 110,000 brains to work for the company—not just 10,000. He inspired 110,000 brains to work for innovation, not just the R&D personnel.

Consequently, Motorola established a process for gathering and evaluating employee ideas. Galvin communicated his expectations that everyone at the company would try to think outside the box. Once innovation started occurring, Galvin himself recognized the innovators and shared the financial rewards from the savings that resulted. Motorola has never done better than during those years. The company was truly a vibrant, growing, and profitable company from 1987 to 1992, when Motorola was committed to applying the Six Sigma methodology rigorously throughout the corporation.

To establish a process for innovation, employees must be inspired for creativity. To be creative, employees must feel comfortable in their jobs. Employees spend most of their active lives at work, and they want to be valuable and know society feels good about the work they do. Too many times, they are discouraged from being creative or innovative; instead, they are given more constraints and discouraged to do anything outside their everyday job requirements.

Business growth requires personal creativity from thinking employees, channeling that creativity to innovation, and converting that innovation to economic gains. Unless there is an established process, growth through innovation cannot be achieved. Innovation can be measured in terms of ideas, recommendations, suggestions, publications, or patents. I still personally thank Motorola for rewarding $100 per published page of articles in any magazine. I personally benefitted from that incentive, as it led me to write more than 100 articles (and now a few books). From a business perspective it is a small incentive; however, it created significant results. Among other contributions, I developed a new test for semiconductor chips as a result of such an incentive.

In *Straight from the CEO* (1998), G. William Dauphinais and Colin Price quantified the degree of creativity required for developing innovative ideas. From a pool of 3000 ideas, about 300 are processed for their innovative potential. About 125 are converted to full-fledged projects, and about 2 products are eventually launched. It becomes obvious, then, that to launch an innovative product, R&D engineers alone cannot do it. On the contrary, launching an innovative product requires the creativity of the entire workforce. J. Morcott Southwood at Dana Corporation observed that the ratio of implemented to suggested ideas improved continually.

As the leadership provides focus and structure for creativity, employee ideas become more relevant to the corporate needs instead of just being personally creative. However, employees' intellectual freedom must be nurtured and nourished to extract great ideas from them. Any one of us has a potential to be a great innovator, given the right environment. Companies that facilitate employees' creativity have realized tremendous financial benefits. Actually, most of the growth of businesses, when it occurs, happens through innovation. Figure 12-7 captures the innovative process in an organization. Accordingly, an organization must establish a system for managing the innovation process effectively and measuring its effectiveness for continually improving the economic value of innovation.

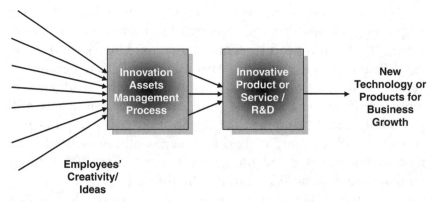

FIGURE 12-7.   Creativity to innovation conversion process.

TAKE AWAY

1. Growth and profitability must be distinctly managed.

2. Profitability depends more on internal factors, while growth depends more on external factors. Growth is realized through strategy, and profits through execution.

3. Without profitability no growth can be sustained, and without growth no business can survive long.

4. The categories that primarily focus on growth are Leadership, Rate of Improvement, Employee Innovation, and Service and Growth.

5. The categories that primarily focus on profit are Sales & Distribution, Purchasing, Operational Execution, and Rate of Improvement.

6. Leadership establishes a process for recognizing employee excellence and extraordinary effort; employees innovate new solutions, products or services; and customer service understands customers' present and future needs.

7. Any organization focusing on selling the products or service at the lowest price alone will not survive long, because some competitors somewhere can always sell a slightly shoddy ones at a little lower price.

LEADERSHIP FOR PERFORMANCE

Economic conditions have changed during the last several decades, especially during the last 20 years. Electronics play a critical role in how we live today. Practically everything we do from work to play has some electronic component in it. The electronics industry has been growing at a phenomenal rate. Growth in electronics led to the explosion of the information age by accelerating the ease and speed with which information can be gathered and communicated. Today we have easy access to almost unimaginable amounts of information about everything in our lives. The Internet has brought the world together in a global village, making it more interdependent and more complex.

LEADERSHIP CHALLENGES

In these exciting as well as nerve-racking times, customer expectations are changing as well. Customers increasingly want "more, better, faster, and cheaper," and businesses work hard to accommodate them. Businesses have changed from body shops to think tanks, well-managed teamwork has shifted into empowered and self-directed work teams, and the role of leadership has changed from managing people to managing minds. And why not? The average corporate profitability happens to be approximately equal to the average use of a person's brainpower, or about 10 percent. In the past, might was right, and the mightiest or the largest one won. In the future, wise is right, and the wisest leaders are going to be successful.

Figure 13-1 shows the changes taking place that affect the corporate world as well as the leadership. Leading in the past looked more like managing a business that was material-dependent, where the cost of goods was significant. Instead, the leadership of today and the future must be prepared to handle more intellectual resources in addition to the material resources.

The recent fallout in the leadership at several large corporations can be attributed to the wrong expectations that society sets for our leaders. The size of the company, growth at any cost, and the financial power of the CEO became the measures of leadership success in the business community. Everyone wants to bootstrap a business with someone else's money. There were times when other people's money was handled with care and returned with interest by businesses. Now Other People's Money, known as OPM (pronounced as the word *opium*), is used as a virtual drug to get the leadership

Organism	Then	Now
Society	Structured family Interdependence Task-oriented Mechanical skills (used material) Sound-driven (hearing senses)	Spontaneous family Independence Systems-oriented Computer skills (use information) Light-driven (visual senses)
Corporation	Quarterly returns driven Acceptable products or services Buy cheapest parts and materials Hire employees for productivity Hierarchical management for tasks Excellence in production Growth or profitability	Customer-focused Highest quality products and services Buy capable parts and materials Attract people for innovation and value Leadership-inspired executive team Excellence in everything Profitability and growth
Team	Task-oriented Mainly in production to solve problems Targeted product problems Chosen members for teams	Goal-oriented Used in development to prevent problems Seeking systemic solutions Qualified members on teams
Leadership	Management-oriented Directing for performance Profitability-focused External measurements	Leadership-driven Inspiring for success Performance-focused Internal measurements

FIGURE 13-1.   Recent organizational dynamics.

intoxicated. The leaders of many dot-coms, Enron, and similar companies took their shareholders' investment for granted, consumed ineffectively, and lost most of it.

Today's epidemic breakdown in business ethics is a manifestation of our deteriorating social fabric, with decreasing family support, increasing gaps in the social safety net, reduced awareness of personal values, and an increasingly cut-throat competitive environment. The difficulties of companies such as Enron really arose because the leaders' compensation was excessively dependent on the stock value, an uncontrollable variable. The incentives offered to these executives were huge—beyond human needs—to the point where running the corporation became a game, not necessarily serious work. Manipulating people, processes, and numbers to show desired results became the norm at such companies. People who highlighted these flaws were sidetracked.

In other words, people forgot about their integrity, honesty, and accountability. Instead, they remembered money, cheating, and miscommunication. The basic tenet of good leadership is to serve the people; instead, leaders had their people serve their hidden personal objectives. Instead of demanding integrity and trust, they demonstrated questionable actions and untrustworthiness. I remember reading a few years ago about one CEO who went to New York to buy paintings worth $1 million to decorate his office and another top executive who sent a bouquet to his wife for thousands of dollars at company expense. Remember the CEO of a not-for-profit charity organization flying to Las Vegas for his pleasure in the corporate jet? When some employees make only minimum wage and have no health benefits while the CEO can make more than $100 million, something is terribly wrong.

Sometimes CEOs who are not successful in meeting the high expectations of stakeholders are released with better packages than they had while working at the company. They wait for a period of time and then are hired at another company with an even better salary. Consequently, there are few incentives for honesty and integrity and few repercussions for incompetence.

LESSONS IN LEADERSHIP

True leaders remember their moments of learning or enlightenment throughout their lives. According to Harris Collinwood in his paper "Leaders Remember the Moments and the People that Shaped Them" (*Harvard Business Review*, 2001), leadership is a personal experience. Leadership is really learned through studying examples of people you revered. In almost everyone's list of revered people, parents, teachers, and coaches rank at the top. Leadership begins at home. Leaders themselves indicate that many lessons they've learned came from their own fathers and mothers. Some of these lessons reveal various traits of leadership, including the following:

Style

Listening and being hungry to learn

Focusing on a few critical things

Displaying positive energy and mannerisms

Making quick decisions

Being willing to ask tough questions

Making good judgments despite personal desires

Demonstrating passion, persistence, and partnerships

Seeking order and harmony

Giving freedom for creativity and inspiration

People Development

Being fair-minded

Surrounding oneself with good people

Recognizing strengths and weaknesses

Building consensus

Bringing people into the process of change

Maximizing employees' well-being, not their comfort

Offering carrot-and-stick feedback

Keeping the good people, not necessarily the likable ones

Excellence

Simple planning
Having a sense of urgency
Having clear and strong values
Being clear about the mission
Sharing a vision of excellence
Demonstrating excellence in everything
Having tough standards of achievement
Enjoying the work more than the results

It is interesting to note that most leaders possess most of the needed skills and practice these lessons. However, sometimes these skills can be applied in excess with adverse results. Sometimes pressures cause leaders to make mistakes. One wrong decision, one poor team member, or the inability to ask tough questions can cause a leader to fail. Just as in a mechanical system, for a leadership system to work well, all parts must be the right ones, be integrated well, and function perfectly. If one part fails, the system fails. If one attribute of the leadership system fails, the leadership fails. Results of a failed leadership can be catastrophic.

Daniel Goleman, Richard Boyatzis, and Annie McKee, in their article, *Primal Leadership* (2001) say that the mood and behaviors of the leadership impact the performance of an organization significantly. According to them, "Emotional leadership is the spark that ignites a company's performance." To maintain that spark for positive energy and performance, a leader can periodically answer some questions (see Figure 13-2). Some leaders do their own performance assessment to gauge their performance, target areas for improvement, and identify action items for growth.

Anthony Robbins says in his book *Unlimited Power* (1997) that if a leader wants to set new standards of performance, the leader can identify a role model to emulate. For example, the role models could be Abraham Lincoln for leadership, Mother Teresa for tranquility, Jack Welch for growth, Ronald Reagan

Questions for Personal Growth	Questions for Professional Growth
Who do I want to be like?	What would I like to accomplish in the short term or long term?
Who am I now?	Where I am now?
How do I get there from here?	How do I get there from here?
How do I make change stick?	How do I maintain?
Who can help me?	What do I need and from whom?

FIGURE 13-2. Process for achieving personal and professional growth.

for communication, Benjamin Franklin for humor, Mary Kay Ash for maximizing women's potential, Robert Galvin for excellence, Oprah Winfrey for addressing social issues, Bill Gates for vision, Steve Jobs for innovation, Carly Florina for leading change, M. K. Gandhi for inspiration, and Martin Luther King for oratory skills. He suggests choosing a role model and then copying that person's behavior and practices continually. Even though a person cannot fully emulate someone else, any progress toward acting as the role model can lead to improved performance.

Leadership for profit and growth requires that a leader be futuristic in thinking, honest, knowledgeable about the business, and trusting. The following seven habits of good leadership can be implemented to well execute sound strategies. Lack of results in most cases is not due to strategy; instead, it is due to lack of effective execution of a strategy. The following is a list of seven practices (seven D's) for good leadership.

SEVEN PRACTICES OF GOOD LEADERSHIP FOR SUPERIOR EXECUTION

PERSONAL PRACTICES

Desire.   For a leader to perform well, the desire to perform must exist. The desire can be created by visualizing the end results and the excitement associated with them. The motivator for creating the desire could be a higher or better cause.

Dedication. Once the desire is there, the leader must be dedicated to fulfilling that desire. Dedication means doing whatever it honorably takes—usually everything.

Discipline. To achieve a major accomplishment, a leader must work hard at it. However, this could create an imbalance in one's personal life. Leaders need to be self-disciplined to ensure a balance between their personal and professional lives. Discipline can come in terms of personal organization, methodology, or consistency.

PROFESSIONAL PRACTICES

Define. To achieve the desired level of profitability and growth, leaders must very clearly define, document, and communicate their objectives.

Develop. Leaders must make a habit of developing a plan to address an issue. Developing a plan allows clarity of thought in action, provides guidelines for execution, eases the delegation of responsibility, and establishes the accountability for performance.

Deploy. Success and failure depend upon the execution of a good plan. Leaders must demonstrate task management and high expectations. Successful leaders stay involved and ask tough questions for results and performance.

Deliver. Leaders deliver the results that were planned and demonstrate them through measurement. Delivery provides leaders an opportunity to give feedback, or give "hugs and kicks," as described by Jack Welch, GE's former CEO.

LEADERSHIP EXPECTATIONS

Leadership's primary responsibilities are inspiration and planning. Inspiration involves motivating people to accomplish a common vision that leadership has established. People are inspired by a leader's knowledge, past successes, personal goodness, love and conviction for the vision, good personal habits, and demonstrated care for other people. The presentation skills vary from leader to leader. The high-energy leader inspires instantly, maybe for a short term, while the low-energy leader inspires slowly, maybe for a long term. The high-energy leader is heard more readily because of body language, while the low-energy leader inspires followers to listen. Inspiration helps a leader in enlisting people with a common mind-set to realize the corporate vision.

To create an inspiring and energizing environment, the leader must establish a credible vision—one that is believable, challenging, worth the employees' commitment, and rewarding and that can have a far-reaching impact on society when realized. Achieving the vision requires defining the core values that will be used to make daily decisions about the actions employees will take. The core values highlight personal guidelines that everyone is expected to follow. Normally, they are about integrity, trust, and mutual respect. A company's attitude toward its activities is then guided by these values.

With the vision, core values, and attitude in place, the leader creates a love for the challenge at hand. Once employees fall in love with the vision and the plan to make it real, they create their passion for it. Passion is reflected in their mood. In other words, the leader must be able to communicate and generate the same feelings in employees as the leader feels. That can happen if the employees buy into the vision, values, plan, reward, and recognition. Passion is manifested in terms of employees' dedication, energy, and infectious enthusiasm. When this happens, employees put their time, their talents, and their hearts into the task. They forget about their time or the hours worked.

To ensure that the vision is realized, the leader must communicate with the team members. The leader must seek feedback about his or her style, strengths, and weaknesses. The leader must also provide feedback about the project progress and the employee contributions toward its success. If the results are not achieved, the leadership must identify areas that need improvement.

Leaders must maintain use of appropriate skills if they are to actually lead the troops. At times, the leader must pamper and care for the employees; at other times, the leader must maintain personal integrity and make tough decisions. People in the leadership position possess most of the leadership skills that enable them to achieve that position. Once they get there, however, they may discover that the demands of the position are impossible to fulfill. It is not that the leader is unqualified or incompetent—instead, it is the set of actions the leader is expected to take that are incongruent with the existing constraints.

The dilemma of leadership is to maintain the ability to discriminate between right and wrong—the ability to make the right decision, to act in a timely manner, to make the tough choices, and to prioritize professional and personal needs. In most cases personal aggrandizing is what leads to the downfall of the leadership. Then the challenge is how a leader can better understand and minimize such problems. Only a strong mind and will can lead to making decisions that are good for the company—even when that means not making the decisions that would benefit the leader personally. Not every decision has to be the most favorable to the leader. To tackle such situations, leaders need to evaluate the impact of their decisions outside their personal sphere of influence and to the systems outside the corporate world.

Every success has its price. Leaders who are extremely successful most likely have paid the price in their personal lives, whether in terms of broken relationships or deteriorating health. These leaders have made the decision to sacrifice their personal lives to achieve professional accomplishments or create greater value. That's a choice leaders make personally. Therefore, the issue is how leaders can balance their personal and professional lives. They must be aware of when they need to put a priority on their professional needs and when they should focus on their personal needs. Neither can be ignored consistently for a long period without detrimental results.

LEADERSHIP ROLE MODELS

To overcome the potential for failure, leaders must be willing to look around for other successful leaders to serve as role models. There are many levels of leadership: personal leaders, parents or family, organizational leaders, national leaders, and spiritual leaders. In today's global environments, because many organizations stretch beyond the boundaries of a single country, the leadership of a global organization is often equivalent to or exceeds national leadership. Some corporate budgets are greater than the budgets of many countries. At this stage of corporate leadership, available role models are limited.

Leadership skills from parents or teachers are learned in the early stages of personal development. The corporate leadership role can be assumed anytime during the life of a leader. Over years of growth, parents and children become peers. The age at which individuals become leaders has, in fact, decreased over time. In the information age, it has been possible on occasion for teenagers to become leaders while parents serve as assistants!

Typically, a corporate leadership role is assumed at a later stage of personal development. At this stage, role models available for corporate leaders are limited, and some leaders turn to spiritual models. Because many leaders lack support, the level of goodness of their actions may determine the success or failure of their leadership at a given time.

As they grow older, corporate leaders must move from greatness to goodness. Greatness is an individual accomplishment, while goodness is a balanced use of many skills and team members. A leader can only enable his or her team to succeed through the goodness that increases with a better understanding of the natural world around us. Many great scientists who have struggled with or could not find an answer to a problem have accepted the supremacy of nature or unknowns. Similarly, when corporate leaders do not have answers to all the questions or solutions to all the problems, they need to maximize the balanced use of skills of employees and achieve success through the greatness of team members.

Leaders must accept the fact that they cannot control everything. Successful leaders realize that and strive toward balance between professional success and personal contentment. With age, most leaders are humbled through life experiences. They recognize the role of luck in their success. Leaders who choose spiritual role models appear to be peaceful, and they do not depend on material abundance. Failure or success does not change them. They lead a balanced life, are respectful of and thankful to others, and attribute their success to everyone, not just themselves.

While not every day will be a balanced day, every month or every quarter can be a balanced one between family and work. If work brings wealth, family brings health. If work brings money, family brings fortune. Leaders need both health from the family and wealth from the profession, in this order. The order may change temporarily; but if the out-of-balance change becomes a pattern, the correct order must be restored.

If we look at the rules and guidance provided by such superior leaders, they speak less, they reflect more, they are dynamic through doing less, and they are level-headed. The leader's measure of success is a combination of professional and personal achievements. The balanced approach is more sustainable than an extreme approach where all the energy is put in one task to its maximum.

To sustain profitability and growth, the corporate system must be balanced between growth and profitability. A business cannot just maximize growth and forget profitability, or maximize profitability and ignore growth. *At the leadership level, the growth comes from inspiration, while the profitability comes from perspiration.* In other words, leaders must work hard as well as smart—just as they expect their employees to do for them.

One way to emulate good leadership is to follow a similar system—the one we learn throughout our lives. Write simple rules for the corporation in terms of vision, values, beliefs, and commands. Provide maximum individual freedom and inspiration, establish accountability for work ethics and performance, thank employees for results, and take care of employees for their hard and smart work. Any corporate sickness must be treated to cure the root cause, not to remove the symptom.

LEADERSHIP CHARACTERISTICS ACCORDING TO THE MBNQA GUIDELINES

The Malcolm Baldrige National Quality Award (MBNQA) has developed guidelines to assess the leadership's performance.

The MBNQA guidelines include the following aspects for assessing the leadership system of a business:

1. Performance evaluation of senior leaders, including the CEO
2. Performance evaluation of the board of directors
3. Leaders' actions to improve their own performance
4. Two-way communication
5. Values, directions, and performance expectations
6. Focused and balanced value to stakeholders
7. Communication of values, directions, and expectations
8. Environment for empowerment, innovation, and agility
9. Organizational and employee learning
10. Legal and ethical behaviors
11. Support and strength for key communities
12. Management accountability
13. Fiscal accountability
14. Independence in internal and external audits
15. Protection of stockholders' and stakeholders' interests
16. Impact of products, services, and operations on society
17. Key compliance measures for surpassing regulatory and legal requirements
18. Key processes, measures, and goals for risks associated with products or services
19. Public concerns regarding products and services
20. Performance and capabilities review
21. Organizational success, competitive performance, and progress against plans
22. Ability to adapt to or address changing needs
23. Review of key performance measures and findings
24. Actions for continual improvement and innovation
25. Involvement of suppliers and partners

LEADERSHIP IMPROVEMENT

An accountability system is only as good as the leader. Leaders need a way to assess for performance, continual learning, and renewal. A sample Executive Personal Assets Inventory Checklist can be found in Figure 13-3. The intent of this

Q #	Question	Yes	No
1	I understand natural behaviors.		
2	I carry positive energy.		
3	I know what success means to me.		
4	I welcome others instead of being welcomed.		
5	I am willing to start over.		
6	I am ready to face the consequences of my actions.		
7	I enjoy every moment of life.		
8	I thank frequently.		
9	I know the purpose for my life.		
10	I learn from the past.		
11	I love everyone.		
12	I am curious.		
13	I know my natural strengths.		
14	I do everything with a passion.		
15	I seek the purpose of every challenge.		
16	I love the critical moments.		
17	I carry the right lesson forward.		
18	I share my resources at work.		
19	I create opportunities for my people.		
20	I honor my appointments.		
21	I am kind and dependable.		
22	I have no expectations for entitlements.		
23	I know my true worth.		
24	I know my time is limited.		
25	I chose the right over the easy in tough times.		
26	I seek win-win solutions.		
27	I serve my employees.		
28	I balance profit and growth.		
29	I prioritize.		
30	I know my needs.		
31	I communicate personally.		
32	I do not have a stiff neck.		
33	I do not get upset easily.		

FIGURE 13-3. Executive personal assets inventory checklist. (*Julian, 2001.*)

inventory is for leaders to review their own strengths and continue to improve. There is no good or bad score.

At the end of the day, the leader can reflect and answer the question, How was the day? He or she can answer using the following criteria:

Mood	Exciting
Knowledge	Learned
Energy	Positive
Goodness	Helped
Results	Productive

Implementing the Six Sigma Business Scorecard requires a total leadership approach. It requires a balance between production and innovation, productivity and creativity, management and leadership, cost and revenue, personal and professional, and profitability and growth. In the world of business, the requirements for leadership have changed over time. *Instead of managing material and labor, the leadership needs to challenge information and minds.* This requires empathetic and thinking leaders who can use objective and subjective measurements for a sustainable value to stakeholders.

The Six Sigma Business Scorecard is a measure of tangible and intangible accomplishments, just like a combination of profits and growth. Similarly, leadership involves a combination of hard and soft skills. Leaders need to find a personally suitable way to maintain a balance between knowledge and goodness, wealth and health, tangibles and intangibles, and profits and growth.

TAKE AWAY

1. The role of leadership has changed from managing people to inspiring minds. The leadership must be prepared to handle more intellectual resources in addition to the material resources.

2. Any progress toward acting as a role model can lead to improved performance.

3. Lack of results in most cases is not due to strategy; instead, it is due to the lack of effective execution of a strategy.

4. The seven D's for good leadership are Desire (to perform), Dedication (to fulfill the desire), Discipline (balance personal and professional lives), Define (objectives), Develop (a plan to achieve objectives), Deploy (the plan), and Deliver (the results).

5. The leadership's primary responsibilities are inspiration and planning. The leader must be able to communicate and generate the same feelings in employees as the leader feels.

6. The leaders must move personally from great to goodness. While greatness is an individual accomplishment, goodness is a balanced use of many skills and team members.

7. At the leadership level, the growth comes from inspiration, while the profitability comes from perspiration.

FOURTEEN

SIX SIGMA BUSINESS SCORECARD VALIDATION

The Six Sigma Business Scorecard was developed to monitor growth and profitability through a set of measurements applicable in today's business environment. Competitive pressures, customer expectations, and the role of knowledge management require that businesses establish practical measurements that add value to the business. The Six Sigma Business Scorecard provides a framework and a set of initial measurements based on my 20 years of research, experience, and observations as an engineer, a manager, a leader, and a consultant working with about 100 business organizations.

The Six Sigma Business Scorecard can be used both internally by businesses for performance improvement and externally as a leading Business Performance Index (BPIn). The index is based on the business's real performance rather than the market's perception of that performance. The BPIn should provide greater confidence in corporate performance because it is based on more complete information instead of on just financial information.

The main differences between the Six Sigma Business Scorecard and other corporate performance measurements are that it requires increased involvement and accountability of the leadership, it emphasizes management's responsibility for achieving a dramatic rate of improvement, and it encourages employees' intellectual involvement for innovation. The Six

Sigma Business Scorecard has been balanced for cost and revenue, growth and profitability, internal and external research and development, objective and subjective measurements, human and material resources, production and nonproduction processes, and management and employees. The measurements are designed to monitor the corporate performance effectively.

Through discussions with professionals in industry and academia and analysis of data for DOW 30 companies using criteria for America's Most Admired Companies at *Fortune* magazine's website, we validated the effectiveness of the Six Sigma Business Scorecard. Accordingly, companies are rated based on eight criteria: innovation, financial soundness, employee talent, use of corporate assets, long-term investment value, social responsibility, quality of management, and quality of products and services. The data were manipulated in a proprietary manner to apply BPIn measurements to the Dow 30 companies using empirical guidelines. In addition to identifying areas for improvement, the intent was to apply the Six Sigma Business Scorecard measurements at the CEO level and determine the corporate Six Sigma level for the Dow 30 companies as a benchmark.

THE 10 BPIn MEASUREMENTS

The 10 BPIn measurements are listed in Figure 14-1. All seven categories of the Six Sigma Business Scorecard are represented in the BPIn measurements. Initially, subjective measurements, such as Employee Recognition and Employee Involvement, appear difficult to measure. After all, the data for such measure-

---

BPIn Measurements

1. Employee Recognition
2. Profitability
3. Rate of Improvement in Process Performance
4. Recommendations per Employee
5. Total Spend/Sales
6. Suppliers' Defect Rate
7. Operational Cycle Time Variance
8. Operational Sigma
9. New Business/Total Sales
10. Customer Satisfaction

FIGURE 14-1.   Ten BPIn measurements.

ments often do not exist in most companies. In addition, many corporations neglect to measure the rate of improvement, even when it could have been measured.

1. Employee Recognition (Percentage of Employees Recognized by CEO)

Employee recognition is public acknowledgment that inspires employees to perform better and find innovative solutions to their daily challenges at work. Employees value recognition from a CEO who is admired at a much higher level than the supervisor. This measurement creates a mechanism for a company leader to interact with employees positively. This leadership recognition may include CEO awards, lunch with the CEO, a day with the CEO, or a banquet with the CEO. The ownership for this measurement resides with the CEO or equivalent leadership. The more the CEO recognizes employees' contributions, the more motivation employees gather to energize themselves and contribute.

To encourage employee recognition from the top leadership, this measure counts the percentage of employees recognized by the top leadership. Many organizations do not measure the extent of employee recognition, considering it trivial. Given its benefits in terms of happier employees, however, this view is misguided. That's why the Six Sigma Business Scorecard requires leadership to establish a process to track employee recognition. Although some businesses may resist doing so at first, the higher business performance that results from the employees' increased intellectual involvement makes it worthwhile. Some CEOs may think employee recognition doesn't contribute to the company's bottom line, not considering how their appreciation of their employees' contributions can lead to significantly more and better efforts that are critical for productivity and innovation.

2. Profitability (Percentage of Net Income)

The profitability measure is a financial gauge of a company's bottom-line performance. The CEO or equivalent representative is directly responsible for profitability, which, in simple terms, is equal to the total revenue minus the total expenses. Profitability is an existing measure for all companies and requires no additional effort to collect.

Being profitable is the final measure of the success of a business. Sometimes business growth is achieved at the cost of profitability. However, without a profitable performance, no business, regardless of its size, can survive very long. The challenge is to sustain profitability over a long period without sacrificing business growth.

3. RATE OF IMPROVEMENT IN PROCESS PERFORMANCE

The rate of improvement measure reflects the latest requirement for a business to stay ahead of the competition. Today, the performance level itself is not as important as the rate of improvement in performance. Conventionally, business leaders have been satisfied to establish a process performance improvement rate of 10 to 15 percent. Because of bureaucracy and inaccuracies in the measurement system, however, employees rarely see any impact when process performance improves only 10 to 15 percent.

In the Six Sigma culture, the rate of improvement in process performance may be planned for as much as 90 percent improvement per year. Given the competitive business environment and the race to win market share, a manager's primary responsibility must be to achieve the rate of improvement goal rather than merely to sustain a department. This aggressive goal for rate of improvement forces management to seek innovative ideas from employees in order to achieve dramatic improvement. This approach mandates continual reengineering of processes for profitability and growth.

4. RECOMMENDATIONS PER EMPLOYEE

Intellectual involvement of employees is a critical measure for the growth of the company. The leadership strives to inspire innovation through offering support and communicating expectations. Leadership provides that support through employee recognition and inspires high expectations by setting an aggressive rate of improvement goal. Many superior companies find ways to encourage innovation. Some companies share the savings reaped from employee recommendations, others provide financial incentives for submitting recommendations, still others offer personal recognition from the executive for participating in the recommendation program, and some distribute recognition and rewards for employees who publish articles or submit an innovation for a patent.

The intent is to maximize the intellectual involvement of employees in improving processes and developing new products or services that will contribute to higher profitability and growth. The Six Sigma Business Scorecard guidelines for this measurement were set to facilitate this process; however, a company may implement a different guideline that better meets its needs in encouraging employee participation. For example, employees can be encouraged to submit a recommendation or two every quarter.

5. TOTAL SPENDING/SALES

The total spending per total sales can be easily measured initially as a ratio of cost of goods sold to revenue. This is an existing measurement that corporations already collect. In many cases, however, this measure is used in its absolute terms instead of reviewing it with respect to revenue. The ratio provides a better indicator when it is monitored in relation to its impact on profitability. The objective is to continually reduce the ratio of the cost of goods sold to revenue.

6. SUPPLIERS' DEFECT RATE

Suppliers are a critical part of a company's success. During the past 15 years, corporations have consolidated their supplier base to reduce variation, but this approach can lead to dependency on a small group of vendors. Instead of simply supplying some parts or services, suppliers have become partners in their customers' success. This increased dependency means that the suppliers' performance must be monitored as an internal process to ensure customer satisfaction.

On the Six Sigma Business Scorecard, the quality of suppliers' parts or services is measured in terms of Sigma through the use of *Defects per Unit* (DPU) for products and *Errors per Order* (EPO) for services. The number of opportunities for error is determined according to the number of parts and processes for products and the number of processes and entries, in financial reports or purchase orders, for services. If the DPU or EPO is not tracked, yield numbers can be used to estimate the DPU from

$$DPU = -\ln(\text{yield})$$

In the absence of any supplier performance data, a business can look into approximating the measure by using the total credit requested from suppliers.

7. Operational Cycle Time Variance

In addition to quality, suppliers' responsiveness in meeting customer requirements is a critical measure of customer satisfaction. The normal measure for monitoring on-time delivery is to compare the actual delivery time to the customer's required delivery time. For internal operations, however, the cycle time measurement tracks various segments of the operation, including the following:

Total cycle time = Time between receiving the customer order and payment

Lead time = Time between placing the order and receiving the shipment

Process cycle time = Total time required to complete a process cycle

Production cycle time = Total time required to receive the material on dock and ship the product

Considering the above descriptions of various cycle times or response times, businesses can establish their own internal goals that will vary from company to company. The Operational Cycle Time Variance is a measure of deviation from the planned cycle time. This measure requires that a business establish internal goals for cycle time reduction similar to the other performance objectives. To achieve total customer satisfaction, an organization must supply what the customer loves to have and deliver it when the customer wants it.

8. Operational Sigma

The operational Sigma measurement requires that each process in an organization be executed well. In other words, each process must demonstrate a measure of goodness. In addition, the complexity of each process must be established in terms of opportunities for error. The measure allows determination of DPU and DPMO at each process. Based on the opportunities for error and the DPMO, the Sigma level can be estimated.

To determine the overall Sigma level, a business adds all its DPUs for individual processes to calculate the total DPU (TDPU):

$$TDPU = DPU_1 + DPU_2 + DPU_3 + \cdots + DPU_n$$

Given the TDPU and the number of opportunities for error the final product or service can have from the customers' perspective, the business can then determine its overall Sigma level. The method of counting the number of opportunities is often a source of much debate, since an increase in the number of opportunities tends to improve the Sigma level. However, this inflation of opportunities for improvement runs counter to the intent of continual improvement. As a matter of fact, the goal should be to reduce the number of opportunities in order to reduce the DPU.

9. New Business/Total Sales

No business can remain profitable without growth, and no business can sustain growth without being profitable. For a business to do well, business growth must be managed as a process and monitored for performance. A growth officer should be appointed to ensure that new opportunities are created for growth. The growth officer's role is to manage internal as well as external research and development processes.

The ratio of new business to total sales is an excellent measure demonstrating growth in new areas. Business growth may come from current customers, but growth due to increasing dependency on one or two customers is prone to volatility because it ties the health of the business directly to the state of its customers' businesses. To reduce the risk of excessive dependence on one customer or a few customers, businesses must set goals to diversify the client base and grow the revenue base. The new business/total sales ratio is a measure of growth and includes the business received for new products or from new customers.

10. CUSTOMER SATISFACTION

Customer satisfaction is a measure of the customers' perception of the quality of products or services they purchase. Surveys are good tools to measure customer satisfaction. Most companies perform their own customer satisfaction surveys. Sometimes third parties, such as trade associations, conduct customer satisfaction surveys within specific industries.

An organization must conduct customer satisfaction surveys to assess customers' future requirements and their experience with the products or services received. Surveys can also be used to create or earn customers' goodwill. Well-designed customer surveys must have questions designed to learn about customers' needs, likes, and preferences—not just about specific product or service requirements. The intent should be to learn about customers' experiences and their willingness to explore a deeper relationship with the company. Businesses should try to solicit this information through various communication media, including the customer survey.

APPLICATION OF BPIn TO DOW 30 COMPANIES

The BPIn measurements were estimated for the Dow 30 companies based on public information as well as estimates of the performance of these companies. The measurements are displayed in Figures 14-2, 14-3, and 14-4.

Figure 14-2 lists all measurements (M1 through M10) and assigned values for the Dow 30 companies arranged in descending order of BPIn, or the corporate wellness index. Company names have been removed to protect confidentiality and privacy of information. The BPIn varies from 53.6 to 84.3. The measurements for Employees' Recognition, Rate of

Improvement, Suppliers' Performance, and Customer Satisfaction were estimated. The other measurements were derived from available online information or annual reports.

In Figure 14-3, BPIn measurements were used to calculate DPU, DPMO, and the Sigma level. The Sigma level varies between 3.06 and 4.02, with a corresponding improvement rate of 90 percent in DPMO. The correlation between profitability and BPIn is about 70 percent, and the correlation between the profitability and DPMO is over 75 percent. The correlation between the DPU and DPMO varies based on organizational complexity, measured in terms of the number of executive decision makers.

Figure 14-4 shows that the median Sigma level is 3.61, which means that 50 percent of the companies are below 3.61 and 50 percent are above. The median DPMO is about

M1	M2	M3	M4	M5	M6	M7	M8	M9	M10	BPIn
10.94	15	15.94	8.06	5	4.9	2.5	3.77	8.08	9.8	84.3
9.6	15	15.42	8.16	5	4.83	5	3.84	7.22	9.66	83.44
11.13	13.5	15.54	7.88	3.75	4.93	4.5	3.78	7.6	9.86	83.04
11.09	15	14.66	7.14	4.5	4.83	4.5	3.24	6.52	9.67	81.48
11.36	13.5	14	7.6	2.5	4.92	4.5	3.95	8.81	9.83	81.23
11.82	13.5	14.6	7.91	2.5	4.89	4.5	3.37	7.7	9.79	81.23
9.95	12	15.48	8.7	2.5	4.93	4.5	4.22	8.6	9.87	80.9
10.92	15	14.36	7.19	4.5	4.85	3.75	3.41	6.42	9.69	80.34
10.28	13.5	15.94	8.29	5	4.86	1.5	3.54	7.38	9.72	80.25
10.91	15	13.96	6.73	4.5	4.88	3.75	3.59	6.36	9.77	79.8
11.69	12	14.96	7.76	2.5	4.92	3.75	3.95	7.64	9.84	79.32
9.32	12	15.84	7.79	2.5	4.88	5	3.04	6.23	9.77	77.28
10.08	13.5	14.44	7.06	1.5	4.88	5	4.35	7.06	9.75	77.15
10.52	15	13.36	6.95	1.5	4.84	5	3.47	6.63	9.68	77.06
9.87	12	15.08	7.77	1.5	4.86	5	3.6	6.85	9.72	76.42
10.5	15	12.78	6.95	1.5	4.84	5	4.38	6.3	9.67	76.12
10.95	9	15.18	7.31	1.5	4.87	5	3.84	7.03	9.74	74.44
10.19	9	15.74	6.97	1.5	4.82	5	4.38	7.7	9.64	74.02
11.42	13.5	13.48	6.7	1.5	4.67	5	3.5	5.04	9.34	73.16
10.22	12	13.2	7	2.5	4.83	2.5	3.57	6.61	9.66	72.06
9.36	12	13.48	6.72	1.5	4.81	5	3.77	6.12	9.62	72.02
9.81	12	12.2	6.59	1.5	4.88	5	3.37	6.33	9.76	71.99
9.95	9	14.48	6.41	1.5	4.91	5	3.24	6.34	9.81	71.53
11.64	4.5	11.06	5.5	1.5	4.87	5	3.89	5.78	9.74	63.42
8.49	15	6.48	3.67	2.5	4.74	5	3.78	4.21	9.47	62.48
9.05	4.5	11.82	6.39	1.5	4.81	5	3.66	5.97	9.61	62.01
9	0	12.84	6.27	1.5	4.84	5	3.55	6.39	9.67	59.1
9.39	4.5	9.96	5.66	1.5	4.79	5	2.52	5.28	9.57	58.89
9.81	0	10.5	5	3.75	4.82	3.75	3.59	5.08	9.65	55.86
9.39	0	10.26	5.17	1.5	4.76	5	4.03	4.96	9.51	53.59

FIGURE 14-2. Estimated BPIn measurements for Dow 30 companies (sorted by BPIn).

Profitability	BPIn	DPU	Executives	DPMO	Sigma
10.82	81.23	0.2078	36	5,773.5	4.02
7.18	77.28	0.2578	39	6,610.1	3.98
4.86	80.9	0.2119	30	7,064.8	3.95
17.64	84.3	0.1708	24	7,267.7	3.94
29.04	83.44	0.181	25	7,239.3	3.94
17.17	80.34	0.2189	22	9,950.5	3.82
8.89	81.23	0.2079	19	10,778	3.79
9.52	83.04	0.1859	17	10,935	3.78
8.99	77.15	0.2594	21	12,352	3.74
3.03	74.02	0.3008	22	13,674	3.7
4.86	71.99	0.3287	23	14,289	3.68
0.57	63.42	0.4554	32	14,230	3.68
6.55	72.06	0.3277	23	14,249	3.68
17.1	76.12	0.2729	16	17,056	3.61
10.87	80.25	0.22	13	16,924	3.61
11	73.16	0.3125	18	17,361	3.6
3.97	74.44	0.2952	16	18,449	3.58
5.68	76.42	0.2689	14	19,207	3.57
5.81	79.32	0.2317	12	19,310	3.56
3.94	71.53	0.3351	17	19,711	3.56
19.75	79.8	0.2257	11	20,515	3.54
15.78	81.48	0.2048	10	20,484	3.54
15.26	77.06	0.2605	12	21,712	3.52
6.95	72.02	0.3282	14	23,871	3.48
−0.63	55.86	0.5822	24	24,260	3.47
0.34	62.01	0.4778	16	29,862	3.37
0.9	58.89	0.5296	16	33,098	3.33
14.68	62.48	0.4704	12	39,198	3.25
−4.57	53.59	0.6238	16	38,986	3.25
−0.42	59.1	0.5259	9	58,435	3.06

FIGURE 14-3. Estimated BPIn and Sigma performance of Dow 30 companies (sorted by Sigma level).

17,000 and DPU is 0.27. The median number of executives making strategic decisions in a large corporation is 17. When a company starts implementing Six Sigma methodology and uses the Six Sigma Business Scorecard to measure its performance, initial measurements provide a benchmark of its performance, which the company can use to develop a plan to improve performance. A Sigma level of 3.61 is close to its expected statistical value before initiating the Six Sigma strategy.

CORRELATION BETWEEN BPIn AND PROFITABILITY PERFORMANCE

Finally, Figure 14-5 demonstrates the correlation between profitability and BPIn. The correlation appears to be strong

Measurements	Average	Range	Median
M1	10.29	3.33	10.2
M2	10.85	15	12
M3	13.57	9.46	14.18
M4	6.91	5.03	6.99
M5	2.53	3.5	2
M6	4.85	0.26	4.85
M7	4.47	3.5	5
M8	3.7	1.86	3.72
M9	6.61	4.6	6.47
M10	9.7	0.53	9.71
DPU	0.3	0.45	0.27
DPMO	19,095	52,662	17,209
Sigma level	3.62	0.96	3.61

FIGURE 14-4. Summary of BPIn measurements for Dow 30 companies.

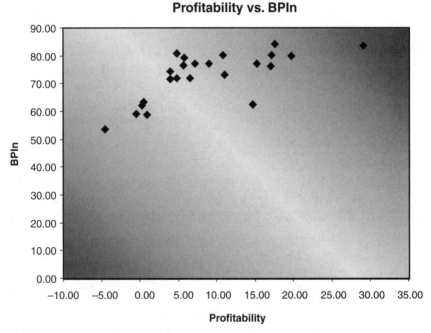

FIGURE 14-5. BPIn and profitability scatter plot.

and should improve over time as the business streamlines its operations and closely monitors its processes for continual improvement. The Six Sigma Business Scorecard measurements used in BPIn include measurements for leadership, managers, employees, sales, purchasing, service, innovation, service, responsiveness, and execution of improvement strategies. The strong correlation between profitability and performance using BPIn proves that BPIn can be used as a leading indicator of the corporate profitability. By improving performance and monitoring with the Six Sigma Business Scorecard measurements and BPIn, leadership can guide the corporation for balanced growth and profitability over time.

The corporate DPMO and the associated Sigma level would be significantly affected if either the number of executives reporting to the CEO or COO or the corporate performance were to change dramatically. Because the executives count cannot be increased much for economic reasons, the focus must be to improve the corporate performance dramatically. The comprehensiveness of the BPIn measurements and the Six Sigma Business Scorecard categories requires improvement in all aspects of business or a balanced approach throughout the corporation. This makes the Six Sigma Business Scorecard a more robust measurement methodology as it is not sensitive to minor variations in corporate performance. The leadership can spend more resources on improving processes rather than let the measurement methods distract from performance. Such improvement is deep-rooted, sustainable, and cost-effective.

The minimum and maximum BPIns appear to be 40 and 99 percent, respectively, based on the Six Sigma Business Scorecard methodology. The 1 percent below the level of perfection is planned as an incentive to improve relationships with customers. The Sigma levels associated with minimum and maximum BPIns depend on the number of executives in a corporation. The best Sigma level for a Dow 30 company could be about 5, and the worst-case Sigma level could be 2.8 without a significant change in operating conditions. However, the Sigma levels could be different under worst or best cases due to variations in the number of executives responsible for making decisions, or representing accountability for corporate failure or success. The primary objective of implementing the BPIn is to continually improve corporate performance and realize maximum benefits for stakeholders.

TAKE AWAY

1. The estimated BPIn for Dow 30 corporations in 2003 ranged from 53.6 to 84.3 percent.

2. The average BPIn for Dow 30 corporations in 2003 was 70 percent, implying a significant opportunity for improvement in business performance.

3. BPIn value of 70 percent is also the minimal value to sustain profits.

4. The Corporate Sigma level varied between 3.06 and 4.02 as a benchmark.

5. Corporations still spend a lot more resources as a percent of sales impacting the net income.

6. Business performance measures must be used as leading indicators for guiding corporate performance up, rather than historical snapshot to judge the acts of leadership.

FIELD
IMPLEMENTATION

The Six Sigma Business Scorecard was developed as a tool to optimize business performance for sustained profitable growth. Its seven elements and ten measurements were identified based on experience working with dozens of organizations representing small to large corporations, addressing concerns with current scorecards, and adapting to the changing business model in knowledge economy. Bob Galvin mentioned during his interview that one should not expect the world switching over to the Six Sigma Business Scorecard, but the world would adapt parts of the Business Scorecard.

Reviewing back 3 years later, readers worldwide have sent emails expressing their liking to the Six Sigma Business Scorecard, and the book in general. Practitioners have asked questions to clarify implementation of the Business Scorecard. I have worked with a company, Ideal Aerosmith, to develop a case study for proving the Business Scorecard model, and gain experience. In the meantime, the book has been referenced in academics and business while dealing with dashboards and scorecards. To date, thousands of readers have been exposed to the Business Scorecard, providing them an alternative to their current methodologies for scorecards, especially the well-known Balanced Scorecard.

The main challenges of current scorecard have been heard everywhere. For example, I asked in one of the business process management conferences in 2006, how many of them have implemented a well-known scorecard. All hands went up. When asked about how many have improved their business using scorecard or like their scorecard, none raised their hand. My understanding is that many companies have implemented scorecards and dashboards to display the data; however, measurements have not been aligned with value stream to impact the top or bottom line. Besides, current scorecards do not focus on human capital utilization, performance improvement, and organizational structure. Today's dashboarding is a fad, and scorecards are fragments unable to steer an organization in the direction for leadership in their industries.

At a macro level, it appears that during the last 15 years, we have seen corporations having fewer innovative products, reduced employee benefits, fewer jobs, and more layoffs. Globalization has impacted every business where competencies have been distributed, instead of being centralized. Internet opened access to human capital worldwide, and networking has allowed formation of virtual teams, and outsourcing has enabled remote business processes. However, the current scorecards' focus is balance instead of relevance of measurements. Balancing measurements has become a breakthrough concept to diversify measurements beyond financials. **Simply balancing measurements is not enough.** Instead, our measurements must be purposeful to drive improvement, innovation, and practice to implement without a significant additional cost, as measurements are part of the process. No measurement must be implemented unless it makes an economic sense, and provides a value proposition through improvement and innovation. Measurements must continually change as new opportunities for improvement are identified. A minimal set of measurements must be used to monitor business performance.

John J. Hovis, Chief Operating Officer of Goal Centrix, Inc., says that an enterprise suffers from a seemingly incurable disease—the inability to execute well. While there are certainly any number of reasons why companies only manage to execute 10 to 20 percent of the strategy they put in place, the common denominator for poor execution seems to be a universal belief by management that "if we simply measure it, it will get done." Our research shows that execution is "management" of activity, not simply the "measurement" of activity. Managers need an activity framework to effectively manage forward, where measurement is an integral part of the process. *Six Sigma Business Scorecard* provides an activity management framework that leverages the structure of Six Sigma thinking and takes it to the next level, the requisite level for execution, by defining the multitude of core activities essential for exceptional enterprise execution, while at the same time providing the rich enterprise-level performance measurement scorecard to complement that activity framework.

Bill Cooper of Lear Corporation has been more direct in his approach to implementing measurements. Accordingly,

> Without value-added metrics and goals that are realistic and stretching for key business processes, the top management is simply "shooting in the dark" with the wishes of hitting a target that their stakeholders could only hope they achieve without crystal balls. Hence, there lies the scenario of 99 percent of all businesses today, unless you too only think with their same philosophy that financial goals are the only targets needed, and ignore the nonfinancial goals.

In order to achieve customer satisfaction, all key customer-oriented processes (COPs) that contribute directly to customer satisfaction must meet their goals, and in order for the COPs to hit their marks, all upstream support-oriented processes (SOPs) must also meet their applicable goals. Along the way, management-oriented processes (MOPs) assess the performance of COPs and SOPs to produce organization-wide decisions, objectives, and/or changes. Hence, a firm is only as strong as their weakest key process, for a process is not usually a key, unless it is value added.

The business scorecard used in tracking processes gives us the opportunity to strategically look at all upstream input processes in relationship to any and all other internal process metrics. Hence, weighing of upstream process metrics, like SOPs and MOPs, gives us the opportunity to predict the final outputs of our COPs, since they are the indicators of customer satisfaction, which are often "real time." The Six Sigma Business Scorecard builds on process-based measurements, establishes their significance, and integrates them into the business model to predict corporate performance.

Lonnie Rogers, President of Ideal Aerosmith, an entrepreneur and seasoned executive, has decided to lead the organization for sustained profitable growth. Aerosmith, Inc. is a leading supplier of precision inertial guidance test systems, rotational rate tables and centrifuges, and high dynamic flight test tables. Our systems test inertial sensors and navigation and guidance systems, and provide dynamic motion simulation for missiles, aircraft, launch vehicles, and automobiles.

Ideal was founded in 1938 to manufacture test equipment for the aviation maintenance industry. Ideal is located in East Grand Forks, Minnesota and has additional facilities in Phoenix, Arizona; Pittsburg, Pennsylvania; and Menlo Park, California. Ideal designs and manufactures motion simulators used to test various navigation systems and motion sensors. Ideal also provide engineering and assembly services for avionics and commercial test equipment.

Ideal employs about 125 employees who have been committed to serve customers, and do whatever is necessary to help the company become a successful company through profitable growth. Ideal had implemented ISO 9001 system, and had heard about Six Sigma through its customers: Rockwell International or Honeywell. Ideal had profitable years that led to a decision to do bigger and better things to serve customers better, and receive more business form them. In other words, higher customer satisfaction and business growth became motivators for implementing Six Sigma. When the market conditions changed, the focus quickly became profitability and growth, in this order of significance.

Ideal's executive team consists of its team of managers headed by its president. The team had an excellent personal rapport, camaraderie, and friendship style of doing things. The friendship style discouraged challenging each other for better performance, a typical scenario, and accommodated each other's weaknesses, as a friendly gesture. When the decision had to be made, president's experience mandated many of them due to lack of effective measurements, and the team struggled in making tough decisions regarding process improvement. In other words, the team experienced a resistance to change the friendship style of management, and create a sense of accountability.

Mr. Rogers had decided to build a system to ensure sustained success beyond his years, and would enable his team to gain valuable management experience for personal and professional growth. However, the challenge being a small company, cost of getting an outside help was a significant one to deal with. Mr. Rogers took the Six Sigma path, as his customers had taken, to improve performance of business processes, and implement business systems. The approach that was taken at Ideal included the following:

1. Assess opportunity to improve financials due to inefficiencies in the operations
2. Practice the Six Sigma methodology
3. Implement the Six Sigma Business Scorecard (because of the author's involvement)

The business opportunity was identified using several tools for assessing current performance. Tools such as ISO 9001 system, McKinsey's 7S (Shared Values, Structure, Systems, Style, Staff, Skills, and Strategy) methodology, the Six Sigma Business Scorecard, and performance analysis using current measurements were deployed. The ISO 9001 allows one to evaluate business processes, the 7S methodology allows evaluation of an organization for its components, the Six Sigma Business Scorecard provides a big picture of performance, and performance analysis quantifies and validates opportunities. Internal survey was conducted to gather employees' perception of the business performance. ISO 9001 has about 25 major elements, 7S has about 7 elements, and the Six Sigma Business Scorecard has 7 elements.

Ideal's business performance was evaluated for the customer, leadership, operations, and supplier performances. The entire organization was assessed using the Six Sigma Business Scorecard. Data was collected from finance, manufacturing, engineering, sales, purchasing, and quality. Key managers were interviewed to understand their operations and challenges.

The qualitative and quantitative data was analyzed using statistical software. The analysis highlighted that Ideal had the following strengths, such as:

1. Product quality was very good due to the experience level and awareness to quality issues.
2. Innovative solutions in developing customized engineering solutions to its customer.
3. Positive employee attitude because of caring and friendly leadership.
4. Strong customer relationships.

Analysis also showed that Ideal faced some challenges. Challenges included heavy workload, fire fighting, rework and repair, parts unavailability due to long lead times, disconnected departmental systems, software errors, and the adverse impact of cost of quality on profitability, and summarized in a report, which was an outcome of the analysis of data and observations collected for the years 1999 to 2003. In order to provide an annualized picture, the performance figures for 2003 are extrapolated based on the available data from the first and/or second quarters of 2003. The final report highlighted significant opportunities for impacting the bottom line, as shown in Figure 15-1. Based on the analysis, three major areas were selected for defining the Six Sigma projects.

Concurrently, the Ideal leadership committed to achieve significant improvement in profitability and growth. In order

Opportunity	Savings potential (% of sales)
1. Increase project efficiency	5%
2. Improve inventory turns	4%
3. Reduce total cost of purchasing	
3a. Supplier consolidation	1.5%
3b. Design for six sigma	1.5%
4. Reduce project cycle time	2%

FIGURE 15-1.   Business opportunity analysis.

to align conflicting priorities, and setting inconsistencies, Ideal's vision, beliefs or values, goals, and initiatives were developed. To develop the corporate vision, all senior level managers were interviewed about their expectation of the organization, their personal style of leading people, their perceptions of Ideal's strengths and weaknesses, their personal and professional growth expectations, and their view of the future state of Ideal. Employee survey was conducted to hear out employees' view of Ideal. Employees highlighted their knowledge of work, support from supervisors, and confidence in Ideal's leadership as strengths of Ideal. Employees also highlighted need to develop shared action plan for company's future direction, recognition, and development of employees.

While developing the company's vision, the leadership team met several times and gained common ground to develop Ideal's common vision, values, goals, and initiatives. There was a discussion about what rate the company should grow. An aggressive rate for growth was thought to be the way to go initially, which was later adjusted to realistic growth rate. The following is a draft of Ideal's vision, beliefs, goals, and initiatives (tactics).

Vision
Leading provider of integrated test systems, engineering, and manufacturing solutions to meet customers' outsourcing needs

Beliefs
Demonstrate respect for everyone
Uncompromising integrity
Maintain a fun and learning work environment

Goals
Work closely with customers
Promote innovation, teamwork, and excellence
Maintain superior financials for profitability and growth

Initiatives
Integrate all Ideal locations seamlessly
Implement the Six Sigma methodology to achieve excellence
Manage projects well for efficiency and profitability

Implement a business scorecard to manage business by facts

Grow business to achieve Ideal's mission

Create a fun and learning environment for employee development and growth

Besides, a mission statement was developed to establish a target for achieving growth and size of the company. Developing vision is a first step in steering the company in the right direction, and making progress. Vision without an action plan sounds like a statement. Thus, the Ideal team sat down right away and developed action plans for these tactics. While developing plans, a owner was designated for each initiative to maintain focus, and for establishing accountability for progress. The action plan must encourage wider participation, and actions must address what action who is going to own, what the target date of completion is, and whose help would be needed to prevent failure. The actions must be detailed enough to ensure specificity, clear enough to communicate, measurably verifiable, and provide roadmap toward the corporate vision. To realize the vision, one must begin with an end in mind, visualize the results, identify steps backward and forward, anticipate problems, find a way around them, challenge employees to bring out its best, and define a target or several incremental targets to guarantee progress.

After defining corporate vision, beliefs, goals, and initiatives, the Six Sigma Business Scorecard was used to assess baseline performance. In absence of a system to continually assess corporate performance, the designated owner had to gather data, estimate when possible, extract information, and assess the performance. From the scorecard, opportunities were identified and initiatives and plans were revisited, and adjusted. While establishing baseline, the challenging measures were CEO recognition of employees, rate of improvement, and employee recommendations. Appropriate action plans were developed in these areas. Due to lack of an existing infrastructure to gather effective information, difficulties were experienced, but the leaders were able to gather information. Throughout, limited outside help was available as

Cat.	Measurement	Feb. 03	Aug. 03	June 06*
LNP	Employee Recognition	65	100	100
LNP	Profitability	33	90	70
MAI	Rate of Improvement	25	0	55
EAI	Employee Recommendations	100	65	25
PSM	CoGS/Revenue	40	52	62
PSM	Suppliers' Sigma	25	37	50
OE	Operational Sigma	25	15	25
OE	Cycle Time variance	80	86	100
SND	% New Business/Total Sales	50	100	100
SAG	Customer Satisfaction	90	80	50
	BPIn	55	61	65

* Estimated snapshot at the author's request.

FIGURE 15-2. Ideal business scorecard.

needed. Figure 15-2 shows Ideal's performance at different times.

Barbara Schultz, then vice president–operations, had led the performance improvement initiative and summarized the implementation of the Business Scorecard as follows:

BUSINESS SCORECARD IMPLEMENTATION STEPS

1. Review and understand the Six Sigma Business Scorecard.

2. Review each measurement in the Business Scorecard with management team to discuss relevancy of each measurement and what it means.

3. Assign vice president–operations, or equivalent, to lead scorecard implementation, a team leader for each measurement.

4. Each team leader will ask two others in company (best to have those that have some control over outcome of measurement) to be on their team for developing an action plan (as shown in Figure 15-3), and implementing the measurement.

Example of an initiative plan

Initiative: **Create a fun and learning environment for employee growth.** **Leader: J.S.**
Objective: Ensure employees' training and development needs are met in a fun and learning environment.
Strategy: Implement employee development plans.

AI #	Action	Responsibility	Committed Completion Date	Critical Resource Required	Status/Comments
1	Form a cross-functional team of representatives from departments.	JS	2/26/03	Limit team size to 4	
2	Introduction meeting-communicate the initiative to the team members.	JS	2/26/03	Getting hold of phoenix personnel	
3	Talk to sample of employees to address the following: 1. Employee feedback 2. Performance 3. Training 4. Development plans 5. Fun environment 6. Employee participation 7. Turnover	Team members	3/7/03	Sample of 12 employees randomly selected to address all locations provide necessary input	Make a form for consistency.

FIGURE 15-3. Sample of an action plan for ideal's initiatives.

5. Each team is responsible for implementing measurement and the team leader reports measurement to vice president–operations monthly.

6. Three measurements were already in place. They were:
   a. Profitability
   b. Spend/sales
   c. Vendor defect rate

7. Other measurements take about 1–2 months to get systems in place to begin measuring. They are:
   a. *Recognition*—We set up two recognition systems. One is Employee of the Quarter. The Employee of the Quarter is recommended by peers and through a systematic review the Employee of the Quarter is determined, and final approval is with the president. We call this *President's Award*. The Employee of the Quarter receives a preferred parking spot, and prizes worth $500. The other recognition is, we call, the Treasure Chest. These are prizes valued under $10 that any employee can give to any other employee, when they feel the employee has gone above and beyond what is expected, or maybe to pick up the employee's spirits.
   b. *Recommendations*—We set up a recommendation system where the employee can email their recommendations or place them in a recommendation box located at each Ideal location. We set up a team to review them monthly and provide feedback to the employee. This team rotates quarterly. Awards may be given based on the quality of the recommendation. Recommendations are tracked as to how many recommendations were implemented, and if feedbacks were provided.
   c. *Lead-time variance*—This measurement took us a while to understand what information we were trying to gather. Once we decided on the information, we had to set up a process in order to track quoted lead-time versus actual lead-time. We used current software to accomplish retrieving needed info for the measurement.
   d. *New business*—We gathered this information using a specific "user field" in our current software, which made it easy to calculate at the end of the month.

8. Once these seven measurements were set up, we started publishing the Business Scorecard and making estimate for the last three. This started 3 months after beginning this process.

9. The last measurements took longer to set up—about 5 months.

    a. *Operational sigma*—We use three key processes to arrive at this number—Production, Quality, and Engineering. The quality defects were already being measured so that the production and the engineering defects had to set up systems to gather the information. Once the systems were set up, the data was easy to arrive at.

    b. *Customer satisfaction*—We came up with a customer satisfaction survey to send to our customer, 60 days after delivery of the product. The sales person determines if a survey should be sent. We use the results of the survey to determine the percentage of customer satisfaction. In addition, we take away 5 percent for every customer complaint during the month, and add 10 percent for every unsolicited "thank you" we receive during the month.

    c. *Rate of improvement*—The last measurement is the continuous improvement measurement. In order to calculate this, we had to have department measurements set up—three from each department. The manager of the department establishes the baseline and set a goal 12–18 months out. This goal is compared to the baseline to determine the percentage of improvement expected over the time frame selected. The percentage of improvement is divided into the number of months, so we know how much improvement is needed each month. This monthly improvement is compared to the goal to determine rate of improvement. All the rates of improvement percentages are averaged to come up with the monthly rate of improvement.

10. The Business Scorecard form is completed monthly with the measurements coming from the team leader. Our goal is to have monthly measurements by 15th of the month,

and the business scorecard by the 20th. The Business Scorecard is posted on the Portal (employee communications software) and bulletin boards. It is presented to the management team monthly, ISO management review meeting quarterly, during the company meeting quarterly and the management quarterly meeting. We track it both monthly and quarterly, and post a graph showing the results to see the trends.

11. Each manager is responsible to review and their measurements monthly, involving others from their department, and will make a presentation to other managers on their measurements and problems to solve. Each manager will be responsible for a presentation every 6 months.

Identifying departmental measures required some learning about the process thinking. Typically, we manage process through its output, which could be numerous, thus making it difficult to measure the process performance effectively. The Six Sigma Business Scorecard emphasizes the process-based measurements. To identify process-based measurements, three questions must be answered for critical processes in each department:

1. What is the purpose of the process? Why does the process exist?
2. What are the outputs and deliverables of the process?
3. What's the measure of goodness of the selected deliverable important to customer satisfaction?

The process analysis is summarized on a form in terms of process steps, its deliverable, opportunities for making mistakes, quality, timeliness, and cost measurements, if applicable, to ensure thoroughness of the process and prioritize based on the economic or business impact. Following is a list of steps each department manages for identifying their measurements, and achieving an aggressive rate of improvement:

1. Identify key processes in each department, define unit, and opportunities for error.

2. Establish effectiveness criteria (define defect or error).

3. Identify measurement of effectiveness.

4. Establish a rate of improvement goal.

5. Develop improvement action plans with clearly defined responsibilities.

6. Establish a procedure to ensure improvement.

7. Collect data and display the trends and levels against goal line.

8. Report performance to operational excellence champion.

At this stage of implementation of the Business Scorecard, manager's process has been harmonized, measurements have been identified, goals have been established, and review process has been practiced. The next question is how to improve a process. For improving processes, department managers attended the Six Sigma Green Belt training. The training used their current projects for improving effectiveness of the training, and realizing return on investment. Meetings were held at a location away from the plant to minimize distractions, and strengthening camaraderie from friendly to professional level. Here, they challenged each other for lack of performance against their common goals. Several new ideas and solutions were developed in this training. As identified in the business opportunity analysis, three projects were selected to improve the business performance. The projects were management of engineering projects, inventory reduction, and engineering process improvement.

Besides, implementation of the scorecard and process improvement, several other actions were taken. President's total commitment to business performance improvement, and speed of organizational changes, and manager's exceptional team work made significant contributions to the success of Ideal's initiative to improve business performance. Employees' embrace of the initiative and support to the leadership, and their participation in the idea recommendation program kept the program going. Celebrations to recognize exceptional employee performance were highly regarded by everyone.

Overall it took Ideal about two months to implement the initial Six Sigma Business Scorecard. Managers were able to start measuring their departments and posted performance measurements electronically on bulletin board that was discussed at company meetings. The measurements were revised as needed to suit Ideal's needs. After one year, we evaluated at each measurement for value and ability to indicate problems and future trends. Then, the Business Scorecard was revised again to specially meet Ideal's needs. As a result of the Business Scorecard, the following benefits were observed:

1. Easy and quick way to get started so that the employees could see Six Sigma in process early on
2. Accountability, checks, and balances
3. Visibility to all employee of our efforts
4. Guidance through setting up measurements because they were preestablished
5. Roadmap and tools to identify areas for improvement
6. Improved teamwork, particularly among managers
7. Combination of Six Sigma and the Business Scorecard led to improvement in profits by 3 percent of sales

In a recent note from Barb Schultz, Barb has mentioned that Ideal has improved its performance. However, the company leadership has made decisions to invest profits back in the company at certain risk to ensure continual leadership in the marketplace. Without significant investment in acquiring new businesses, Ideal would have been quite profitable company. According to Barb,

> We have done many things right because of Six Sigma and the Business Scorecard implementation. We are all now accountable and rely on measurements to make decisions. This has been a real advantage. Communicating to our employees and within the management group has been a great advantage, along with employee recognition. We have chosen to focus on some areas of the Six Sigma Business Scorecard, instead of all areas at this time, as other processes have been fully implemented, and due to our strategy for growth.

TAKE AWAY

Ideal's experience demonstrates that in order to sustain business performance, one must do the following:

1. The Six Sigma Business Scorecard is more focused on execution of strategic intent through effective measurements rather than developing strategies at the management level alone.
2. Instead of strategy maps, one must deploy strategy for execution maps to ensure effective implementation of sound strategies. Without execution map, sound strategy can become a "sorrow" strategy.
3. One must keep the implementation of the Business Scorecard simple, and do not let it become a project in itself.
4. Today, we have maximum measurements with minimal effectiveness. We must minimize measurements, but maximize benefits from the "right" measurements.
5. Excellent implementation of the Business Scorecard requires simple maps, clear roles, and involvement of employees.
6. One must be able to change measurements as necessary; however, maintain use of a consistent framework to ensure performance improvement, and omission of critical business processes.
7. Any business scorecard implementation must be linked to strategies related to sustained profitable growth at a given time.

INTEGRATING THE SIX SIGMA BUSINESS SCORECARD AND QUALITY MANAGEMENT SYSTEMS

The ISO 9000 standards have been practiced worldwide for about 15 years. Based on common standards of performance, they were created to harmonize trade among countries. The standards included the best practices commonly used in successful organizations. The standards were released in 1987 and first revised in 1994. The main purpose of the 1994 revisions was to the clarify requirements. The second revision, the ISO 9001:2000 standards (released at the end of 2000), incorporates more comprehensive changes based on the experience gained through worldwide implementation.

The intent of the ISO 9000 standards has stayed the same since their inception—to promote quality thinking and utilize best practices throughout organizations. They enable a business to implement effective business processes so it can provide products or services that meet customer-specified requirements. The quality thinking requires that each process be managed such that it performs as well as needed to meet customer requirements. To know how well a process is performing, the business must verify its performance at appropriate points throughout operations. Then if a product or service does not meet the customer requirements, it can investigate causes of nonconformity and initiate appropriate corrective action.

To implement such an accountability system consistently worldwide, the International Organization of Standardization (ISO) released the standards through its national bodies of member countries. The registration architecture is shown in Figure 16-1. As the diagram indicates, several organizations play roles in ensuring effective implementation of the standards. Challenges are experienced at every level—with the most significant one being the auditor level. The auditor must ensure that the requirements of the standards were implemented effectively and complied with consistently. Given that assessment of compliance is easier than assessment of

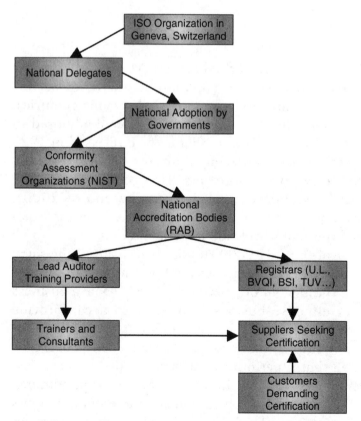

FIGURE 16-1.   Organizational structure for ISO 9001 certification.

effectiveness, auditors often focus on compliance more heavily than on effectiveness.

Accordingly, corporations seeking ISO certification adopted the mantra of consistency—*Do what you say, and say what you do*—but changed it to mean *Document what you do, irrespective of its effectiveness*. Though incorrect in principle, that is what happened. As a result, organizations saw little measurable benefit of the ISO 9000 system on their bottom line. When the ISO 9001:2000 standards were revised in 2000, they were revised to promote effectiveness over compliance.

In order to improve compliance, ISO has incorporated use of Deming's PDCA (Plan, Do, Check, Act) model. The intent of PDCA to ensure enough product is planned to meet customer requirements, work is done with the parts as necessary, the output is checked to ensure satisfactory performance, and necessary actions are taken to address deviant conditions. The implication of using PDCA is that checking is built in every process for ensuring acceptable performance against the specified limits. Practice of checking against limits eliminated the need for establishing performance targets, and without performance targets, one will face challenges in achieving the Six Sigma performance. To reduce the problem of checking, a new process management model has been developed, which is called the 4P model of process management where the four Ps represent Prepare, Perform, Perfect, and Progress. The current version of ISO 9001 is based on the PDCA model in which the input is customer requirements and the deliverable is the process output meeting customer requirements. The concept of requirements has been understood as specifications or tolerances. Processes are designed to produce products within these limits, and verification is limited to the established requirements or the tolerance.

The ISO 9001:2000 version defines PDCA as follows:

- **Plan:** Establish the objectives and processes necessary to deliver results in accordance with customer requirements and the organization's policies.
- **Do:** Implement the process.

- **Check:** Monitor and measure processes and products against policies, objectives and requirements for the product, and report the results.
- **Act:** Take actions to continually improve the process.

Looking into an underlying source of problems in current manufacturing facilities—the culture of fighting fires, acceptability mind-set, and ignored variability in product performance—we need to move away from the concept of process control to the concept of process management. Process management implies establishing expected target performance, providing necessary resources, ensuring superior execution of a sound methodology, and taking care of exceptions with appropriate process controls. Applying this understanding of process management to PDCA, one can see gaps.

The 4P cycle, as shown in Figure 16-2, incorporates the wisdom of Shewhart, Ishikawa, Juran, Taguchi, and Deming. Accordingly,

- **Prepare** represents ensuring good inputs to the process. The inputs consist of Ishikawa's 4Ms (material, machines, methods, and manpower or people). The objective is to ensure these four Ms are received well as inputs to ensure the process output will be on target.

FIGURE 16-2.   4Ps of process management.

- **Perform** implies the process steps are well defined, and understood for effective and consistent execution.
- **Perfect** means assessing the performance against the established target performance. If the process output is not on target, the gap is assessed.
- **Progress** leads to reducing variability or the gap around the target, and striving toward the targeted performance.

By continually applying the 4P cycle, one can reengineer the process to achieve the results desired by the customer through better process management, instead of increased inspection. Benefits of applying the 4P cycle of process management are as follows:

1. Better quality products or services, and happier customers
2. Better process knowledge, thus thinking workers
3. Deviation points to the source, rather than blaming the operator
4. A significant improvement in performance
5. Less emphasis on testing, and more emphasis on designed quality

Apart from the ISO 9000 standards, the QS-9000 and the Malcolm Baldrige Award guidelines included requirements for customer satisfaction and continual improvement. Those two requirements have been added to the ISO 9001:2000 standards. The ISO 9001 standards now include effectiveness measurements, continual improvement, data analysis, and customer satisfaction.

The quality management system requirements of ISO 9001:2000 state, "The organization shall establish, document, implement, and maintain a quality management system and continually improve its effectiveness in accordance with the requirements of this International Standard."

To accomplish the above requirements, the activities include

- Identifying the processes needed for the quality management system
- Establishing the sequence of these processes

- Determining criteria and methods to ensure effectiveness of these processes
- Ensuring the availability of information necessary for monitoring these processes
- Measuring and analyzing these processes
- Implementing actions to achieve continual improvement of these processes
- Ensuring control over outsourced processes

Based on the measurements of the quality management processes, ISO 9001:2000 requirements assert that the organization shall analyze appropriate data to demonstrate the suitability and effectiveness of the quality management system, as well as to determine opportunities for continual improvement of the effectiveness of the quality management system.

SIX SIGMA METHODOLOGY FOR CONTINUAL IMPROVEMENT

Six Sigma uses the DMAIC methodology for dramatic improvement. The DMAIC methodology consists of five steps: Define, Measure, Analyze, Improve, and Control. It uses tools such as Kano's quality chart to highlight customers' real expectations as well as stakeholders' analysis to align resources in the theater of operations. The DMAIC methodology uses additional tools such as the SIPOC chart (Suppliers, Inputs, Process, Outputs, and Customers), multivary analysis to isolate process problems, design of experiments to improve, control charts to maintain, and C_p and C_{pk} to measure capability (C_p is a measure of process capability, and C_{pk} is a measure of process performance).

The DMAIC methodology can be used to fulfill the continual improvement part of the ISO 9001 requirements. The objective is to establish a baseline to define opportunities for improvement, measure process effectiveness to quantify the magnitude of the problems, analyze data to investigate the root causes of the problems, apply necessary experimental techniques to improve the process, and monitor techniques to sustain the process effectiveness and improvement.

SIX SIGMA BUSINESS SCORECARD FOR EFFECTIVENESS

The Six Sigma Business Scorecard consists of measurements for various leadership and operational processes specified as quality management processes in the ISO 9001:2000 standards. To establish measurements for effectiveness that meet ISO 9001 requirements, it is necessary to construct a business process flowchart, identify key processes, establish criteria for effectiveness, and determine measurements for effectiveness. The Six Sigma Business Scorecard consists of the seven categories listed below:

1. Leadership and profitability
2. Process management and improvement
3. Employee innovation
4. Purchasing and supplier management
5. Operational excellence
6. Sales and distribution
7. Service and growth

The corresponding 10 key measurements include

1. Employee recognition
2. Profitability
3. Rate of improvement in process performance
4. Recommendations per employee
5. Total spending/sales
6. Suppliers' defect rate
7. Operational cycle time variance
8. Operational sigma
9. New business/total sales
10. Customer satisfaction

These measurements focus on purchasing, sales, customer satisfaction, rate of improvement, operational excellence, innovation, and profitability (the ultimate measure of business effectiveness). The operational excellence category includes the remaining key business processes. The Six Sigma Business Scorecard can therefore be a great mechanism for implementing effectiveness measures in a quality management system and can make the ISO 9001:2000 a value-added system by improving the corporate performance.

INTEGRATING THE SIX SIGMA BUSINESS SCORECARD WITH ISO 9001

To integrate Six Sigma and ISO 9001 requirements, business leaders must understand how the ISO 9001 standard on performance will affect various elements of the business. The Six Sigma Business Scorecard can be used to identify key business processes, establish measures of performance, and link those measures to the profitability of the company. As required by the ISO 9001 standard, therefore, a procedure must be established to list a final set of measurements, collect necessary data, analyze data, identify opportunities for improvement, and formalize improvement action using the corrective action system. Once the measurements are established, reviews of performance levels and trends become part of the management review meeting. During this review meeting, performance against planned goals is discussed, and any necessary actions to achieve continual improvement goals are implemented.

To benefit from the Six Sigma methodology and comply with the continual improvement requirements of the ISO 9001 standards, businesses must apply the methodology economically. Improvements to products or services need to be linked to corporate profitability. The improvements must be accomplished in a cost-effective manner and only if benefits outweigh the cost. In addition, businesses need to be sure they have a process for sustaining improvements.

COST CONTAINMENT

The Six Sigma Business Scorecard attempts to address concerns about implementing Six Sigma economically. The methodology requires a top-down approach driven by the corporate strategy. When the top-down approach is combined with the tactical approach of Six Sigma, powerful results are achieved. To achieve the real improvement possible with the Six Sigma Business Scorecard, one of the first tasks that needs to be undertaken is a business opportunity analysis. This will enable the business to identify the best projects to tackle for improved profitability. The tangible benefits realized by the company may include reduced waste, lower reprocessing and warranty costs, increased margins and hence improved profitability, and lower investment in working capital. In addition, intangible benefits can be gained through an ingrained culture of customer and quality focus.

Currently, the Breakthrough strategy of implementing Six Sigma requires a significant initial expense for training corporatewide Executives, Champions, Green Belts, and Black Belts in Six Sigma awareness and methodology. Companies that have committed to significant training corporatewide without proper planning struggle to identify projects that will allow them to achieve any return on the investment. The goal of Six Sigma is to achieve the highest quality at the lowest cost. The costs of implementing the Breakthrough methodology are estimated to break down in the following way: communication (1 percent), recognition (4 percent), planning (5 percent), organizational restructuring (10 percent), project implementation (30 percent), and training (50 percent).

THE IMPROVEMENT METHODOLOGY

The Six Sigma improvement methodology incorporates a review of strategic alignment and an analysis of the benefits and costs involved, which addresses some of the concerns mentioned above. The six-step approach is shown in Figure 16-3.

Step 1: Perform Opportunity Analysis. As discussed earlier, this critical step is often overlooked. Improving quality adds some costs—test equipment, inspections, improvements to processes, higher-quality materials, and so on. To ensure that the quality improvement initiative will add value to an organization, comparing the additional costs to the benefits to be gained is critical. Only if benefits outweigh the cost should the initiative

FIGURE 16-3.   Six Sigma improvement methodology.

be pursued. The benefits typically include lower costs due to less scrap and fewer errors, defects, or returns; increased revenues due to being able to charge a higher price for a superior product or service; or revenue growth through increased sales.

To make this step a periodic activity, businesses are encouraged to conduct quarterly analyses of data from a variety of sources. They might look at customer feedback, internal audit findings, corrective and preventive action analysis, internal rejects, capability analysis, repair and rework data, suppliers' performance, innovation, rate of improvement, and employee feedback. Such analysis can help the business to reprioritize opportunities for improvement and realign the company with business objectives. Some of the process improvement projects may very well be considered preventive actions. ISO 9001 standards require that organizations take preventive actions. The business opportunity analysis must identify projects that will lead to an increase in profitability by improving processes and reducing wastes in the operations that eat up the profitability.

Step 2: Establish Goals. The key to successfully achieving continual improvement is the ability to establish business performance goals and communicate clearly to everyone involved. Using a Business Scorecard framework, management identifies goals in each area. While most companies continue to give financial objectives much greater weight than any other areas, meeting goals in all areas is extremely important for sustained financial success. In the ISO 9001:2000 framework, continual improvement goals are set for products and processes. Therefore, companies that use ISO 9001 must set goals that involve processes and product characteristics that will enhance customer satisfaction and lead to reduction in waste.

Step 3: Ensure Cross-functional Buy-in. The next step is to align the operational initiatives to the departmental objectives. The management team must look at the impact the objectives will have on various departments, how different departments will be involved, and the resources required to accomplish the continual improvement goals. The role of Six Sigma gets defined at this point. Whether it is implemented with the intent to reduce cost (increase profits), enhance customer satisfaction, or optimize process efficiency, the Six Sigma methodology can be adapted to play a role and help achieve desired results.

As Six Sigma is integrated in the ISO 9001:2000 infrastructure, awareness training for all employees must be conducted to ensure a common understanding of the revised standards' new emphasis and the role of Six Sigma methodology. The amount of training needed is a big concern for many companies. Based on the success of various Six Sigma initiatives across the industries, Green Belt training supported by a few in-house or outsourced Black Belts is the way for most companies to proceed. Most problems can be solved using the tools taught in Green Belt training.

A Green Belt program, one to two weeks long, enables participants to understand and apply Six Sigma concepts. The Green Belt program is designed to teach the DMAIC methodology. The course incorporates the practical statistical and graphical tools to improve processes. Typically, the Green Belt curriculum includes the following:

Green Belt Training Contents

Understanding Six Sigma

Six Sigma Methodology

> Define
> Measure
> Analyze
> Improve
> Control

Six Sigma Measurements

Implementing Six Sigma

Roles to Play

> Champion
> Black Belt
> Green Belt

Monitoring Project Progress

Step 4: Analyze and Improve.   Once you have screened the initiatives to include only the ones that show the promise of increasing shareholder value, the next step is to perform a detailed analysis of the problem. This step leverages the robust analytical framework offered by the Six Sigma methodology to achieve breakthrough results. Additional activities include establishing short- and long-term performance measures and improvement targets.

Simple tools, such as the cause-and-effect diagram, regression analysis, multivary charts, and Pareto analysis, can be used to look for root causes of the problem. If necessary, experiments can get to the bottom of the problem. Once the causes are identified, the possible process changes are examined based on process knowledge and experience. The revised process is validated before implementing changes. Many companies make the mistake of overlooking the validation step in the rush to implement the process changes. When making changes to processes, businesses must look at various approaches beyond current practices in order to develop a breakthrough solution. In other words, some innovation must occur in creating the new process in order to realize dramatic improvement in the process output.

Step 5: Manage Change.   Step 5 involves planning to ease the organization's transition to the new performance levels. This step also involves identifying the resources required to make the change and considering how the change will be introduced with the least amount of disruption to current operations. It involves detailed evaluation of changes required across five key dimensions—material, methods, machines, skills, and technology. Employees will need training to successfully implement the new processes.

Step 6: Enhance and Sustain Performance.   Many initiatives meet their demise as they become one-time improvement programs. The company's performance tends to return to its previous levels once the focus on improvement is removed. The best way to keep process improvement continual is to monitor process performance accurately and in a timely manner. In addition, recognizing successes is another way of sustaining the rate of improvement. The growth officer should utilize a companywide database to record process improvement activities, lessons learned, tools used, and results achieved. Sharing this information freely with everyone can create new ideas, facilitate learning for growth, spark new solutions, and create new value propositions.

STEPS TO INTEGRATE SIX SIGMA AND ISO 9001

To integrate Six Sigma into a company's ISO 9001 quality management system, a corporatewide initiative must be established for positive impact on the bottom line. The following steps can provide some guidance for integrating Six Sigma and ISO 9000:

1. Identify opportunities for improvement based on an analysis of data from purchasing, customers, processes, internal audits, corrective actions, and management reviews.
2. Commit to applying the Six Sigma methodology through the ISO 9000 system.

3. Create a vision for the Six Sigma initiative.

4. Revise the quality policy to demonstrate commitment to continual improvement using the Six Sigma methodology.

5. Identify procedures that will be affected by the integration of Six Sigma and an ISO 9001 quality management system. This may include Management Review, Continual Improvement, Corrective Action, and the Quality Manual.

6. Establish a training plan for all employees, including the leadership, to make sure everyone is aware of the ongoing initiative.

7. Determine the need for Green Belt and Black Belt training based on the projects or opportunities that have been selected for action.

8. Select individuals for the appropriate training. Normally Green Belt training lasts one to two weeks, while Black Belt training may last two to five weeks. In either case participants are advised to identify current projects on which to work.

9. Hire a training firm to facilitate the learning. The focus of the training is more on application than on skill development.

10. Hire a proven mentor with hands-on experience in successful problem solving—someone who is skilled in using the various tools and problem-solving methods in the Black Belt curriculum. Look for someone with the leadership skills necessary to bring out the best innovative solutions for dramatic improvement.

11. Assign very clear responsibilities for improvement and successful implementation of process changes.

12. Recognize teams and individuals who brought about dramatic improvement and achieved excellence.

13. Consider promoting people who demonstrably produce value or who show success to growth positions in the company.

14. Monitor progress in the management review meetings, and sustain the improved results and the Six Sigma methodology for new projects.

CONDUCTING GOOD INTERNAL AUDITS

Besides establishing measurements and goals for dramatic improvement, the key quality management process that must exist in the ISO 9001:2000 paradigm is auditing first for effectiveness and then for compliance (instead of auditing just for compliance). To assess implementation of ISO 9001:2000 in an organization, the lead auditors, as well as the internal auditors, must look into both effectiveness and compliance.

The compliance aspect of the audits gained priority early on because the excessive emphasis on documentation made it easy to check. Having a priority on compliance keeps the audit specific, objective, and unchallenged. However, the ISO 9001:2000 standards' focus on effectiveness, reduced emphasis on documentation, and increased leverage for the auditee gives auditors a new challenge. To conduct audits that will truly be useful to the organization, the auditor must be experienced in performing audits well, proven in working in his or her chosen industry, and motivated to make a positive difference.

Good auditors are driven to make a difference in their auditing jobs. They learn quickly to analyze many different processes for their inputs, methods, outputs, and intent of use. Auditors must also be able to ask questions about how the user is trying to improve the process in terms of cost, cycle time, and quality. After systematically assessing the compliance of a process in terms of inputs, process steps, and outputs, the auditor must ask questions that relate to the soundness of the process (i.e., how well the process works).

Does the process produce the desired results? If it does not, the auditor must be able to probe further to guide the operator or the user in seeing what has gone wrong. If the auditor can help a user to understand the process better, identify disconnects in the process, and recognize opportunities for changing the process for the better, then the auditor has performed a good job. The interesting point is that in such audits, the auditee is the one who discovers noncompliances (with the auditor's help, of course).

BENEFITS OF INTEGRATING ISO 9001 AND SIX SIGMA

- Forces leadership to make the ISO 9001 quality management system a tool for improving the profitability
- Reorients the leadership in understanding the ISO 9001 system as a bureaucratic paper monster to a value-added tool for running the business
- Promotes dramatic process improvement results
- Builds on previously learned tools
- Links performance to profitability, thus creating opportunities for sharing benefits with employees
- Forces leadership and employees to think outside the box and to set aggressive goals
- Requires innovation for achieving planned results
- Creates excitement about the corporate quality initiative instead of a test to pass for achieving certification
- Boosts customer perception of the company, as Six Sigma integrated with ISO 9001 is designed to produce superior results and better customer service

The time has come to look at ISO 9001 or an equivalent quality management system from the right perspective—as a system promoting quality thinking in order to manage the business for improving performance. The intent was lost in the race to achieve a certificate, which led to frustration over insignificant changes in performance. Companies adopted the path of least resistance, and auditors conducted audits with minimum aggravation for fear of losing clients or to gain more clients by being an easy registrar.

Companies developed procedures to record what workers did to achieve compliance to the standards' requirements, instead creating procedures that outlined what workers should be doing to achieve expected results. Even internal auditors conducted audits using standard checklists and found no non-compliances against the simple procedures they were auditing.

In other words, ISO 9001 was implemented without its intent, and this led to stagnant quality management systems that companies longed to abandon. Such systems became boring and burdensome to sustain.

With the continual improvement requirement added in the revised version of the ISO standards, the intent to create value is now explicitly stated. Six Sigma can facilitate the value intent of ISO 9001:2000. Companies that are implementing Six Sigma independently of the ISO 9001 infrastructure are finding themselves wasting resources with redundant processes. They are finding it difficult to maintain their Six Sigma initiatives. Teaming up with Six Sigma and the ISO 9001:2000 standards is the right choice.

TAKE AWAY

1. ISO 9000 standards were created to harmonize trade among countries with the intent of promoting quality thinking.

2. The quality thinking requires that each process be managed such that it performs well for meeting business objectives.

3. The Business Scorecard strives for excellence, while standards require acceptance.

4. In order to achieve excellence, corporations must utilize the 4P model of process management that requires to set clear targets of performance.

5. We must produce to established targets for excellence, and deliver to customer within their acceptable limits, preferably on the target.

6. Business management systems such as ISO 9001 require implementation of measures of effectiveness, and methodology to improve.

7. The Business Scorecard can be used to monitor key business processes linking to the profit stream, and Six Sigma can be used as a methodology to improve processes.

8. The Business Scorecard can drive the performance in conjunction with the business management system.

REFERENCES

Bossidy, Larry, and Ram Charan. *Execution: The Discipline of Getting Things Done.* New York: Crown Business, 2002.

Boyd, Ty. *Visions From the Leaders of Today for the Leaders of Tomorrow.* Charlotte, N.C.: Alexa Press, 1991.

Breyfogle III, Forest W. *Implementing Six Sigma: Smarter Solutions Using Statistical Methods.* New York: Wiley, 1999.

Chowdhury, Subir. *The Power of Six Sigma.* Chicago: Dearborn Trade Pub., 2001.

Collins, Jim. *Good to Great: Why Some Companies Make the Leap...and Others Don't.* New York: Harper Business, 2001.

Covey, Stephen R. *Principle-Centered Leadership.* New York: Simon & Schuster, 1991.

Cribbin, James C. *Leadership, Your Competitive Edge.* New York, AMACOM, 1984.

Dauphinais, G. William, and Colin Price (eds.). *Straight From the CEO: The World's Top Business Leaders Reveal Ideas That Every Manager Can Use.* New York: Simon & Schuster, 1998.

Framework for Improving Performance. Oakbrook Terrace, IL. Joint Commission on Accreditation of Healthcare Organizations, 1994.

George, Michael L. *Lean Six Sigma: Combining Six Sigma Quality with Lean Speed.* New York: McGraw-Hill, 2002.

Goldratt, Eliyahu M., and Jeff Cox. *The Goal: A Process for Ongoing Improvement.* Great Barrington, Vt.: North River Press, 1992.

Goleman, Daniel, Richard Boyatzis, and Annie McKee. Primal Leadership. In *Harvard Business Review on Breakthrough Leadership.* Boston: Harvard Business School Press, 2001.

Gomes, Helio. *Quality Quotes.* Milwaukee: ASQC Quality Press, 1996.

Graham, Jacqueline D., and Michael J. Cleary (eds.). *Problem Solving and Planning Tools: Practical Tools for Continuous Improvement,* vol. 2. Miamisburg, Ohio: PQ Systems, Inc., 1992–2000.

Graham, Jacqueline D., and Michael J. Cleary (eds.). *Statistical Tools: Practical Tools for Continuous Improvement,* vol. 1. Miamisburg, Ohio: PQ Systems, Inc., 1992–2000.

Haines, Stephen G. *The Manager's Pocket Guide to Systems Thinking and Learning.* Amherst, Mass.: HRD Press, 1998.

Hamel, Gary, and C. K. Prahalad. *Competing for the Future.* Boston: Harvard Business School Press, 1994.

Hammer, Michael, and Steven A. Stanton. *The Reengineering Revolution.* New York: Harper-Collins Publishers, Inc., 1995.

Harrington, H. James, and Kenneth C. Lomax. *Performance Improvement Methods: Fighting the War on Waste.* New York: McGraw-Hill, 2000.

Harry, Mikel, and Richard Schroeder. *Six Sigma: The Breakthrough Management Strategy.* New York: Doubleday, 2000.

Harvard Business Review on Breakthrough Leadership. Boston: Harvard Business School Press, 2001.

Harvard Business Review on Knowledge Management. Boston: Harvard Business School Press, 1998.

Harvard Business Review on Measuring Corporate Performance. Boston: Harvard Business School Press, 1998.

Hill, Napoleon. *Think and Grow Rich.* New York: Fawcett Crest Books, 1960.

Insights and Inspirations: How Businesses Succeed. Hartford, Conn.: Mass Mutual, 1999.

Jones, Laurie Beth. *Jesus in Blue Jeans: A Practical Guide to Everyday Spirituality.* New York: Hyperion, 1997.

Julian, Larry. *God Is My CEO.* Avon: Adams Media Corporation, 2001.

Juran, J. M. *Quality Control Handbook,* 4th ed. New York: McGraw-Hill, 1988.

Kaplan, Robert S., and David P. Norton. *The Balanced Scorecard.* Boston: Harvard Business School Press, 1996.

Kaplan, Robert S., and David P. Norton. *The Strategy-Focused Organization.* Boston: Harvard Business School Press, 2001.

Klauser, Henriette Anne. *Write It Down, Make It Happen: Knowing What You Want—And Getting It!* New York: Simon & Schuster, 2000.

Kotter, John P. *Leading Change.* Boston: Harvard Business School Press, 1996.

Moore, Geoffrey A. *Crossing the Chasm: Marketing and Selling High-Tech Products to Mainstream Customers.* New York: Harper Business, 1999.

Naumann, Earl, and Steven H. Hoisington, *Customer Centered Six Sigma: Linking Customers, Process Improvement and Financial Results.* Milwaukee, Minn.: ASQ Quality Press, 2001.

Newman, Bill. *10 Exciting Keys to Success: The Power of a Successful Life.* Toowong, Australia: Bill Newman International, 1995.

Pande, Peter S., Robert P. Neuman, and Roland R. Cavanagh. *The Six Sigma Way: An Implementation Guide for Process Improvement Teams.* New York: McGraw-Hill, 2002.

Pyzdek, Thomas. *The Six Sigma Handbook: A Complete Guide for Greenbelts, Blackbelts, and Managers at All Levels.* Tucson, Ariz.: McGraw-Hill, 2001.

Ries, Al. *Focus: The Future of Your Company Depends on It.* New York: Harper Business, 1996.

Robbins, Anthony. *Unlimited Power.* New York: Fireside, 1997.

Strawser, Cornelia J. *Business Statistics of the United States,* 7th ed. Laham, Md.: Bernan, 2001.

Tichy, Noel M. *The Leadership Engine.* New York: Harper Business, 2002.

W. B. Freeman Concepts, Inc. *God's Little Devotional Book on Success.* Tulsa, Okla.: Honor Books, Inc., 1997.

Waitley, Dennis. *Empires of the Mind: Lessons to Lead and Success in a Knowledge-Based World.* New York: William Morrow, 1995.

Womack, James P., and Daniel T. Jones, *Lean Thinking: Banish Waste and Create Wealth in Your Corporation.* New York: Simon & Schuster, 1996.

NOTE FROM THE AUTHOR

Praveen Gupta, president of Accelper Consulting, has developed the following tools:

1. Business Performance Index (BPIn)
2. Corporate Sigma Level Method
3. Innovation Index and Breakthrough Innovation Methodology
4. 4P Model of Process Management
5. Optimized Six Sigma Methodology

In order to help readers and their corporations achieve their fundamental strategy cost effectively, Mr. Gupta and his team offer additional tools and guidance. If you hit an impasse in realizing your corporate vision, and feel like talking to someone who has experienced a lot, you may want to ask Mr. Gupta and his team for their guidance. He has created the following opportunities to interact with his readers:

Discussion Forum: Readers can join their peers on *www.businessscorecard.org* and ask questions about the performance measurements, dashboards, or scorecards. Subject matter experts will be able to answer your questions.

Strategic Planning: If you need to get started using the Six Sigma Business Scorecard, Mr. Gupta's team will be able to look into operations and help you plan for a simple and meaningful implementation of the scorecard.

Training and Consulting: Mr. Gupta's company, Accelper Consulting, can provide help on demand through its services in identifying practical measures leading to BPIn, training in the Six Sigma Business Scorecard, and providing support for improving corporate performance.

Contact Information:

Email: info@accelper.com

Tel: (847) 884-1900

Website: *www.accelper.com*

INDEX